ENVIRONMENTAL REMEDI[illegible] TECHNOLOGIES,
REGULATIONS [illegible]

EARLY DETECTION OF FOREST FIRES FROM SPACE

ENVIRONMENTAL REMEDIATION TECHNOLOGIES, REGULATIONS AND SAFETY

Additional books in this series can be found on Nova's website under the Series tab.

Additional E-books in this series can be found on Nova's website under the E-book tab.

SPACE SCIENCE, EXPLORATION AND POLICIES

Additional books in this series can be found on Nova's website under the Series tab.

Additional E-books in this series can be found on Nova's website under the E-book tab.

ENVIRONMENTAL REMEDIATION TECHNOLOGIES,
REGULATIONS AND SAFETY

EARLY DETECTION OF FOREST FIRES FROM SPACE

G.G. MATVIENKO
S.V. AFONIN
AND
V.V. BELOV

Nova Science Publishers, Inc.
New York

For permission to use material from this book please contact us:
Telephone 631-231-7269; Fax 631-231-8175
Web Site: http://www.novapublishers.com

LIBRARY OF CONGRESS CATALOGING-IN-PUBLICATION DATA

Matvienko, G. G. (Gennadii Grigor evich)
Early detection of forest fires from space / authors, G.G. Matvienko, S.V. Afonin, V.V. Belov.
p. cm.
Includes index.
ISBN 978-1-61324-509-5 (softcover)
1. Forest fires--Remote sensing. I. Afonin, S. V. (Sergei Vasil evich) II. Belov, V. V. (Vladimir Vasil evich) III. Title.
SD421.375.M38 2011
634.9'618--dc23

2011017304

Published by Nova Science Publishers, Inc. † New York

CONTENTS

PREFACE

This book considers the results of the theoretical and practical works dealing with forest fire detection from space; they are performed in Zuev Institute of Atmospheric Optics (IAO), SB RAS, since 1997. The chapter consists of four parts.

Its first part addresses the results of forest fire detection on the territory of Tomsk region for period of 1998-2008 with application of AVHRR/NOAA satellite system. The authors studied the question of early detection of small-sized fires and analyzed the dependence of fire detection results on the time of the monitoring. The detection results, obtained in IAO with the help of the AVHRR/NOAA regional algorithm, are compared with those based on the MODIS Fire Products (MOD14) global algorithm, used for MODIS/EOS system.

The second part of the book presents the methodic foundations of RTM approach to the multispectral monitoring of the earth's surface. The authors analyzed the influences of the distorting effect of the atmosphere and uncertainties of the specification of the atmospheric optical and meteorological parameters on the results of retrieval of the Earth's surface temperature from space.

The third and fourth parts describe the software for implementation of the RTM approach and the results of its practical application.

INTRODUCTION

In remote sensing of the Earth's surface from space, the actual problem of real-time detection of fires in forests and industrial objects is solved. Obviously, it is important to detect a fire at early stages (when its area is less than 5–10 ha), while its extinguishing does not require great efforts. In this case, reliable algorithms, automatically detecting small-sized high-temperature objects (HTO) with area less than 0.1% of the pixel size are required.

An analysis of the algorithms for the fire detection from space published in the literature has allowed us to make the following conclusions. Most fire detection algorithms, used in practice, employ the decision rule $P\{x\} > dP$, where dP is the threshold value of the function $P\{x\}$, and its parameters $\{x\}$ are satellite measurements of albedos and brightness temperatures (or of their functions). The threshold values dP are fixed or can be determined based on statistical analysis of $\{x\}$ in the vicinity of a potential fire. However, the actual parameters of the atmosphere during satellite measurements are in fact disregarded by algorithms used in practice.

Efficient solution of fast control of the environmental state from space is possible only through the application of high-accuracy algorithms of thematic processing and atmospheric correction of space-borne measurements. This is particularly important in temperature sensing of the environment under complex observational conditions including the detection of weak-intensity fires at early stage of their development. Unfortunately, the presently available methods do not solve this problem.

The correction of satellite IR measurements for the distorting effect of the atmosphere with the use of information on the atmospheric state (meteorological and aerosol parameters of the atmosphere) and on the geometry of observations during satellite measurements is undoubtedly important for obtaining maximum accuracy.

The main objectives of this chapter are:

1) To outline a procedure of real-time monitoring of forest fires at the IAO SB RAS with the use of the AVHRR data.
2) To analyze the results of satellite monitoring of boreal forest fires in the Tomsk Region in 1998–2008; to investigate the dependence of the results of fire monitoring on the time of day; and to explore the possibility of early fire detection from satellites.
3) To develop the multispectral RTM approach for monitoring of the earth's surface from space and for detecting the small-sized forest fires; and to perform method's testing in the practice.

Chapter 1

MONITORING OF BOREAL FOREST FIRES IN THE TOMSK REGION OF WESTERN SIBERIA

Real-time detection and monitoring of forest fires in vast and not easily accessible territories of Siberia and Far East are urgent problems for Russia. In 1998-2008, according to the data of the Tomsk Forest Protection Services, 3205 forest fires with a total area of more than 213 000 ha burnt in boreal forests in the Tomsk Region of Western Siberia (55°N to 60°N, 75°E to 90°E, see the map in figure 1).

In the last decade, satellite monitoring of forest fires (SMFF) based, as a rule, on the AVHRR/NOAA data has been widely used in Russia. Since 1996, the National Forest Fire Centre of Russia has had an Internet site (http://nffc.infospace.ru) in which the satellite sensor data on forest fires for the most part of the territory of Russia are daily updated in fire-hazardous seasons.

In addition to the National Forest Fire Centre of Russia, there are some independent regional centers of satellite monitoring of forest fires (in Krasnoyarsk, Irkutsk, Novosibirsk, Tomsk, and Yakutsk). In our opinion, they significantly increase the efficiency of fire detection because of knowledge of specific conditions of monitoring on local territories and implementation of satellite sensor data processing algorithms adapted to these conditions. It is also obvious that regional centers of satellite monitoring can maintain continuous contact with Forest Protection Services.

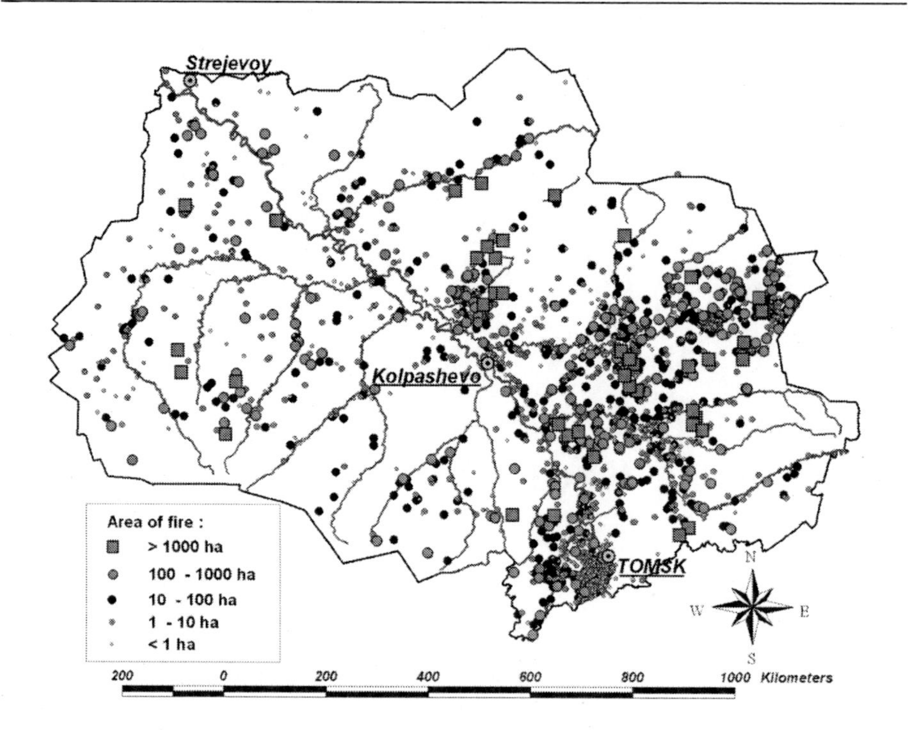

Figure 1. Map of forest fires detected by Forest Protection Services in the Tomsk Region in 1998–2008 (the largest rivers are mapped).

Since 1998, the Institute of Atmospheric Optics (IAO) of the SB RAS has carried out real-time satellite monitoring of forest fires on the territory of Tomsk Region. The Institute has all the required components to solve this problem, such as:

- ScanEx station (http://scanex.ss.msu.ru) for the acquisition of digital satellite sensor data,
- knowledge of the theory of image transfer through the atmosphere and methods of correction of satellite measurements for the distorting effect of the atmosphere,
- long-term practical experience in satellite sensor data interpretation,
- and standard and original software packages for satellite sensor data processing (see, e.g., [[1]-[5]]).

To develop a system of real-time satellite monitoring of forest fires, we first analyze the experience accumulated in Russia and abroad. The majority of works published in the literature contain data on fire monitoring at middle and southern latitudes (Africa, Brazil, Spain, the USA, Canada and Northern Europe). To a lesser degree such investigations were carried out in the Russian boreal forest zone.

In addition, only a few papers addressed the efficiency of early detection of forest fires (EDFF) from satellites. The term *"efficiency of early detection of forest fires"* can be treated in two ways. First, it can be considered as the probability of detecting a small fire with an area of about 1 ha and smaller. This probability, in fact, represents the accuracy of satellite detection of fires at the initial stage of fire burning. Second, this term can be defined as the difference $\Delta T = T_1 - T_2$ between the time of the first satellite fire detection (T_1) and that of its detection by a Forest Protection Services (T_2). Then, the value of ΔT characterizes an advance ($\Delta T > 0$) or a delay ($\Delta T < 0$) of satellite systems of fire monitoring relative to the conventional (ground-based or airborne) systems of fire detection. This characteristic allows the role of satellite sensor data in routine operation of Forest Protection Services to be determined in fire-hazardous seasons (at least as a rough approximation).

1.1. Image Processing

Satellite monitoring of boreal forest fires in the Tomsk Region is carried out every year in the fire-hazardous season from April (May) through September. The procedure of monitoring includes three main stages (described in [[6]] in detail).

1) Reception of digital data from NOAA-12 to NOAA-19 satellites. Pre-processing of satellite sensor data, including calculations of geographic position based on the well-known SGP4 program and data from the Two-Line Elements Set available via Internet source (http://celestrak.com/NORAD/elements/noaa.txt). Refinement of geographic position in semi-automatic operation using reference points and contour hydrographic lines by a special program that allows the geographic position error to be reduced down to 1–2 km.
2) The AVHRR data processing by two algorithms for automatic detection of hotspots on the underlying surface. Automatic rejection of sun glints in satellite images. The results of automatic satellite image processing and the quality of false alarms rejection are

controlled by an operator. To improve the reliability of detecting small fires, a special computer program for automatic comparison of previous satellite images is used. This program identifies the presence of the fire being detected in these images.

3) Generating a map of the Tomsk Region based on the results of satellite monitoring of fires (a fragment of the map is exemplified in figure 2). The map is then delivered to the Forest Protection Services with a text file describing some characteristics of the hotspots detected.

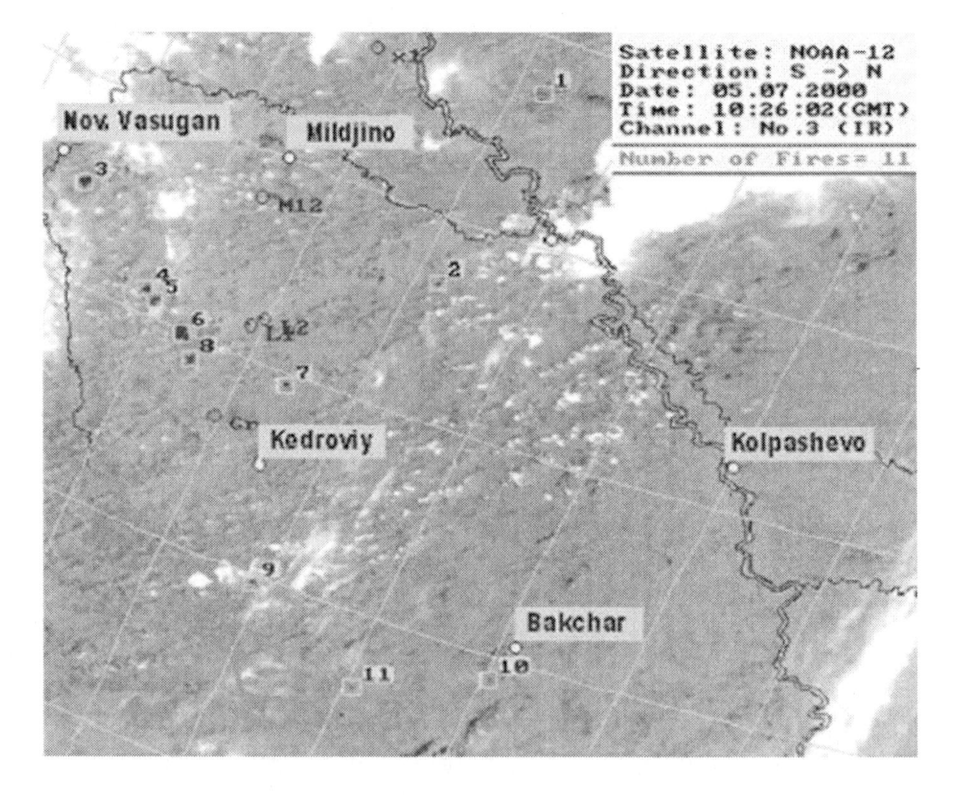

Figure 2. Fragment of the map with the results of NOAA–12 satellite monitoring of forest fires in the Tomsk Region at 18:26 LT on July 5, 2000. The fires detected are marked by small squares and numbered.

Fire Detection Algorithms

According to the available literature data [[8]-[13]] two types of satellite algorithms – Fixed Threshold Techniques and Spatial Analysis Techniques – are mainly used to solve the problem of automatic detection of forest fires. Having tested various threshold algorithms [[5]], in the initial stage of fire monitoring, we chose an algorithm best adapted to the Siberian conditions. It was the algorithm based on [[14]] and developed at the Centre of Space-borne Monitoring at the Institute of Solar-Terrestrial Physics (ISTP) of the SB RAS [[15]].

This algorithm uses four fixed thresholds for the brightness temperature in the AVHRR band 3 ($\mathbf{T_3}$) depending on the albedo in the AVHRR band 1 ($\mathbf{A_1}$):

1) $\mathbf{T_3} > 290$ K for $\mathbf{A_1} < 1$ (and $T_4 > 265$ K),
2) $\mathbf{T_3} > 306$ K for $\mathbf{A_1} < 4$,
3) $\mathbf{T_3} > 316$ K for $\mathbf{A_1} < 10$,
4) $\mathbf{T_3} > 320$ K for $\mathbf{A_1} < 25$ (and $T_4 > 265$K),

given that the rule $\mathbf{T_3 - [T_4 + 3 \cdot (T_4 - T_5)] > 4}$ is fulfilled, where $\mathbf{T_4}$ and $\mathbf{T_5}$ are the brightness temperatures in the AVHRR bands 4 and 5, respectively.

When testing the threshold algorithms for boreal forests on the territory of Tomsk Region with the use of the actual satellite sensor data for 1998 [[5]], the best results were obtained for the ISTP algorithm. At the same time, the results of the testing indicated a need for improving the satellite algorithms for detecting forest fires in the case of monitoring under unfavorable conditions of broken clouds, low-intensity fires, and sun glints from the water surface and clouds. For this purpose, the original Spatial Analysis Algorithm [[16]] was developed at the IAO SB RAS in 1999. The algorithm allowed the contribution of solar radiation in the 3rd AVHRR channel to be considered and the reliability of fire detection under unfavorable conditions of measurements to be increased. It was tested for the satellite sensor data in 1998-2000 and showed high efficiency, versatility, and reliability of its operation.

Thus, since 1999 we have used two algorithms for fire detection from satellites: the basic IAO algorithm and the backup ISTP algorithm. We believe that this approach to monitoring with the use of two different algorithms increases the reliability of the results of fire monitoring where fires are detected by both algorithms. In addition, while the basic algorithm offers advantages over the backup one, our experience shows that in some cases, the latter provides additional data on seats of fires, thereby improving the results of implementation of the former.

Problem of Sun Glints

It is well known that rivers, lakes, wetland, and cloudiness produce sun glints in satellite images when the geometry of observations is unfavorable, and these glints may cause false alarms in monitoring of forest fires. In the Tomsk Region, there are hundreds of rivers (the largest rivers are mapped in figure 1), hundreds of lakes, and large wetland areas, and the average amount of clouds over this territory in the fire-hazardous season (May–September) is about 50%. Therefore, the problem of sun glints is very serious. This fact is supported by the data recorded on June 19, 2000 and tabulated below (table 1).

Table 1. Number of hotspots in satellite images recorded on June 19, 2000

NOAA satellites	Image LT	Sun elevation, deg	**Number of hotspots**			
			T_3>305 K	T_3>310 K	T_3>315 K	T_3>320 K
15	10:10	33.9	3407	263	2	0
14	17:03	44.9	8533	333	7	4
12	17:50	39.4	11765	4387	2206	996
15	20:01	22.6	681	17	2	0

Notes:
LT is the local time of recording of the satellite images (LT=GMT+8),
$\mathbf{T_3}$ is the brightness temperature in the AVHRR band 3.

On that day, the number of hotspots increased by several thousands within a very short period (between 17:03 and 17:50) due to sun glints. The frequency of glint occurrence in satellite images of this region is 5–10% of the number of all images. These situations are most typical of postmeridian images (more than 70% of all images).

The literature [[8],[9],[12]] describes the following criteria for sun glint rejecting: (1) the angle between the hotspot-to-sensor vector and the specular reflection vector is small (~10°) and (2) albedos measured in visible channels exceed the preset hotspot threshold (~0.3). An analysis of the map in figure 1 suggests that most forest fires burnt in the immediate proximity of rivers or other water reservoirs. Another important circumstance follows from the statistics of our measurements: more than 75% of false alarms caused by sun glints in satellite images are from 1 to 3 pixels in size. Consequently, for boreal forests in the Tomsk Region, we must not only solve the problem of sun glint rejections but also distinguish between neighboring small-size sun glints and fires in satellite images. Our experience demonstrates that in our case, the use of the above-listed criteria alone does not solve these problems with sufficient accuracy. Therefore, we use a more complex procedure of automatic rejection of sun glints including:

1) specialized spatial analysis of satellite measurements,
2) threshold values and albedos A_1 and A_2 as well as their relationships determined for the sensor–hotspot–sun geometry of the Tomsk Region,
3) comparison with a previous image in which sun glints are absent,
4) hydrographic data of an electronic map of boreal forests in the Tomsk Region.

Practical application of this procedure demonstrated its high (close to 100%) efficiency.

1.2. Results

The results of satellite monitoring of boreal forests in the Tomsk Region performed at the IAO SB RAS in 1998–2008 were analyzed using specially developed software (described in [[7]] in detail). Fire location data from the Tomsk Forest Protection Services were used as reference data.

To compare the satellite sensor data with the reference data, the following characteristics of fires were used:

- geographic coordinates;
- date and time of fire detection ($\mathbf{T}_{DET}$), localization ($\mathbf{T}_{LOC}$), and extinguishing ($\mathbf{T}_{EXT}$);
- fire areas at the time of their detection ($\mathbf{S}_{DET}$) and extinguishing ($\mathbf{S}_{EXT}$).

To estimate the efficiency of early detection of forest fires from satellites, the fire duration was defined as $[\mathbf{T}_0, \mathbf{T}_{EXT}]$, where $\mathbf{T}_0 = \mathbf{T}_{DET} - \Delta\mathbf{T}$ ($\Delta\mathbf{T}$ depends on $\mathbf{S}_{DET}$). These characteristics are illustrated by figure 3 that shows the following statistical data for 1998–2008: (a) histograms of fire distributions over the areas $\mathbf{S}_{EXT}$ (figure 3a) and histograms of fire distributions over the duration [$\mathbf{T}_{DET}$, $\mathbf{T}_{EXT}$] (figure 3b).

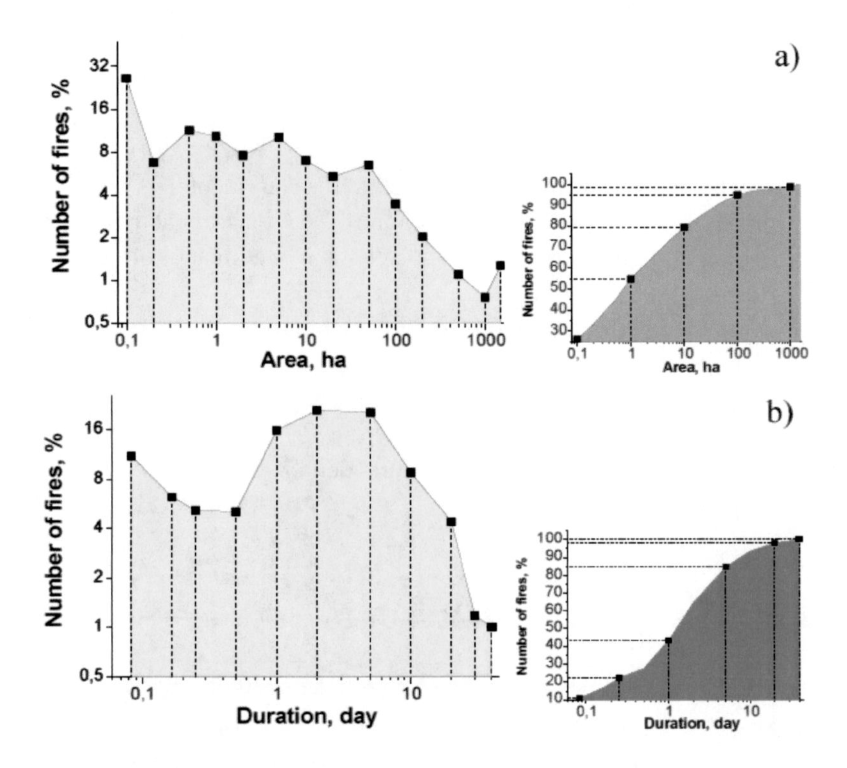

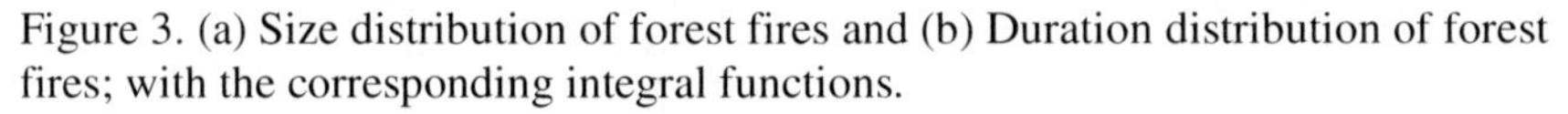

Figure 3. (a) Size distribution of forest fires and (b) Duration distribution of forest fires; with the corresponding integral functions.

It follows that small-size fires with areas less than 1 ha, on the average, account for about 53% of all fires, and fires with short duration (< 24 h) account, on average, account for more than 48% of all fires. Thus, low probabilities of detection from satellites could be expected for more than 50-60% of all fires.

The following procedure was executed with the use of these characteristics for each satellite image:

1) hotspots detected from satellites were mapped;
2) all fires, which actually burnt or potentially might burn in the region under study at the time of monitoring $\mathbf{T}_{IMG}$ (i.e., at $\mathbf{T}_0 \leq \mathbf{T}_{IMG} \leq \mathbf{T}_{EXT}$), were mapped;

3) these two maps were compared, and the pairs of objects that coincided in accordance with the preset spatial criterion (separation of less than 10 km) were selected with allowance for their spatial size and probability of occurrence of a fire in the image set to unity in the interval [$\mathbf{T}_{DET}$, $\mathbf{T}_{LOC}$] and vanishing as $\mathbf{T}_{IMG} \rightarrow \mathbf{T}_0$ or $\mathbf{T}_{IMG} \rightarrow T_{EXT}$.

First, let us examine table 2, where the results of satellite monitoring of boreal forest fires in the Tomsk Region are given for each month of 1998–2000. Table 2 includes the following data:

a) the number of fires detected by the Forest Protection Services ($\mathbf{NF}_{\Sigma}$);
b) the monthly average cloud cover amount, i.e., part of the territory inaccessible for satellite monitoring;
c) the SMFF efficiency ($\mathbf{NF}_{DET}$), i.e., the number of forest fires detected from satellites;
d) the EDFF efficiency ($\mathbf{NF}_{ED}$), i.e., the number of forest fires detected from satellites before they were detected by the Forest Protection Services.

Table 2. Results of satellite monitoring of forest fires in 1998–2000

Month	May	June	July	August	September	**Total**
1998						
NF_{Σ}	24	32	175	147	—	378(-87)
Cloudiness,%	34.6	49.3	20.1	46.4	—	
NF_{DET}	6	2	38	51	—	97
NF_{ED}	5	1	11	25	—	42
1999						
NF_{Σ}	60	31	199	132	60	482(-80)
Cloudiness,%	45.7	55.6	28.5	49.6	61.8	
NF_{DET}	39	9	72	51	22	193
NF_{ED}	20	0	23	33	10	86
2000						
NF_{Σ}	47	93	110	49	19	318(-57)
Cloudiness,%	59.0	35.5	46.9	48.6	70.4	
NF_{DET}	16	25	26	8	2	77
NF_{ED}	6	18	9	1	0	34

It follows from table 2 that the SMFF relative efficiency is in the range 24–40% in different years. A total of 367 sites of fire were detected from satellites in fire-dangerous seasons of 1998–2000, that is, the probability of fire detection was $\mathbf{PF}_{DET}$=31%. The probability of early fire detection **(**$\mathbf{PF}_{ED}$**)** was in the range 11–18% (162 sites), which on average was about 14% for 3 years.

When estimating the SMFF results, it should be kept in mind that some fires (or observation conditions) had the characteristics that made their detection from satellites hardly possible:

a. when the fire duration was so short that there were no images in the time interval [$\mathbf{T}_{DET}$, $\mathbf{T}_{EXT}$],
b. synchronous fires were separated by less than 2 km, i.e., they could not be distinguished in satellite images,
c. the fire was covered by dense clouds during [$\mathbf{T}_{DET}$, $\mathbf{T}_{EXT}$].

The number of such fires for each year (15–25% of the number of all fires) is indicated in parentheses in the last column of table 2. This value seems to be underestimated (for example, this follows from the fact that forest crowns, masking small fires, were present in 96% of cases, but were ignored here). After the fires undetectable from satellites were subtracted from the number of all fires, the average efficiency of satellite monitoring $\mathbf{PF}_{ОБН}$ increased to 39%, and the average efficiency of early fire detection $\mathbf{PF}_{ED}$ increased to 17%. It should be noted that virtually the same probability values were retained for the entire examined period (1998–2008): $\mathbf{PF}_{DET}$=38% and $\mathbf{PF}_{ED}$=18%.

We now will analyze the dependence of the satellite forest fire detection results on the fire size. For this, we turn to figures 5 and 6 which show the histograms of the dependences of the total SMFF efficiency $\mathbf{PF}_{ED}$ versus the fire area $\mathbf{S}_{EXT}$ (figure 4a) and the efficiency of early fire detection versus the fire area $\mathbf{S}_{DET}$ (figure 4b).

According to figure 4a, the efficiency of satellite monitoring as a function of the fire area is: up to 20% for $\mathbf{S}_{EXT}$ less than 1 ha, 40–50% for $\mathbf{S}_{EXT}$≈5–10 ha, and greater than 80% for $\mathbf{S}_{EXT}$ > 50 ha. The minimum area of forest fires detectable from satellites is about 0.1–0.2 ha, and the probability of detecting these fires is about 10%.

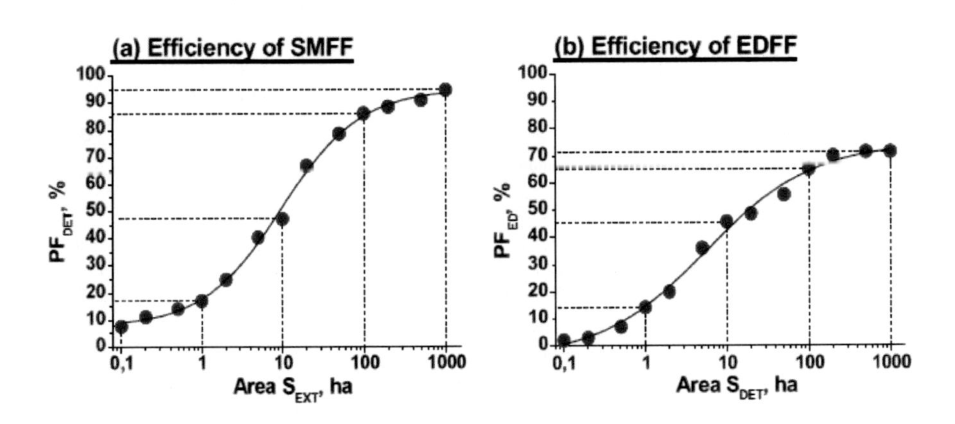

Figure 4. (a) Efficiency of SMFF and (b) efficiency of EDFF.

The experience accumulated by Forest Protection Services demonstrates that a near-ground fire with an area less than 5 ha can be extinguished with high probability. Turning to figure 4b, we can notice fairly high (>15%) efficiency of early detection of fires already at $\mathbf{S}_{DET}\approx1$ ha, increasing to 35–45% for $\mathbf{S}_{DET}\approx5$ ha. Thus, in our case the satellite sensor data can be used efficiently for early detection of fires, when their extinguishing is not very expensive.

Imaging Time-of-Day Effect

In this section, we examine the information content of satellite images for monitoring forest fires as a function of the time of day. To this end, all the SMFF data were divided according to the type of satellite (NOAA-12, -14, -15) and orbit (antemeridian and postmeridian). The urgency of the problem stems from the fact that some monitoring procedures used in Russia are based only on postmeridian images or even on a single daytime image. This approach is motivated by the evident circumstance that postmeridian images are most informative and close in time to breaking-out fires and spreading of already burning ones.

Table 3. Results of satellite monitoring of forest fires as functions of the time of day

Orbits	Antemeridian (a.m.)			Postmeridian (p.m.)			Total
Satellites	NOAA-12	NOAA-14	NOAA-15	NOAA-12	NOAA-14	NOAA-15	
1998							
LT	08:14–10:02	04:56–06:50	–	18:28–20:12	15:08–16:47	–	
N_{IMG}	91	93	–	86	83	–	
$\mathbf{PF}_{DET}$,%	24.7	34.0	–	76.3	80.4	–	94.8%
$\mathbf{PF}_{ED}$,%	9.3	23.3	–	34.9	72.1	–	85.7%
1999							
LT	07:52–09:37	05:39–07:30	–	18:04–19:49	15:44–17:36	20:01–21:40	
N_{IMG}	150	148	–	153	152	153	
$\mathbf{PF}_{DET}$,%	22.2	23.7	–	74.7	67.0	57.7	96.4%
$\mathbf{PF}_{ED}$,%	19.8	14.0	–	54.7	52.3	48.8	91.9%
2000							
LT	07:31–09:45	06:21–09:00	09:51–12:01	17:44–19:45	16:20–18:20	19:59–21:54	
N_{IMG}	121	132	81	126	141	83	
$\mathbf{PF}_{DET}$,%	13.0	15.6	26.0	66.2	75.3	46.8	94.8%
$\mathbf{PF}_{ED}$,%	5.9	0.0	32.4	55.9	55.9	35.3	85.3%

Notes:

LT is the local time of satellite-based image recording (LT=GMT+8), $\mathbf{N}_{IMG}$ is the number of satellite images, the total SMFF results ($\mathbf{PF}_{DET}$) and the efficiency of early detection ($\mathbf{PF}_{ED}$) are given in per cent of the number of all fires for a year (see Table 2).

The data in table 3 demonstrate the SMFF results as functions of the time of day. Antemeridian images have relatively low efficiency, whereas postmeridian images are characterized by much (a factor of 2–3) efficiency. However, from table 3 it is clear that the use of only one (even the most informative) image reduces the efficiency of satellite fire monitoring by 20–25%, and the efficiency of early detection decreases by a factor of 1.5–2. The last column in table 3 gives the total efficiency of satellite monitoring with the use of only postmeridian images. It follows from these data that about 95% of the SMFF results and more than 85% of the early detection results have been using only the postmeridian images. Thus, even though the information content of postmeridian images is high, some useful satellite sensor data are lost due to rejection of antemeridian images, reducing substantially the efficiency of early fire detection. Consequently, only a full SMFF procedure including all satellite images provides a maximum SMFF efficiency and real-time monitoring of the fire dynamics and state of cloudiness in the site of the fire.

Comparison with MODIS Fire Product (MOD14)

Another important element of estimating the efficiency of the technology of real-time forest fire detection from space developed at the IAO SB RAS is its comparison with other algorithms widely used in other centers of satellite monitoring. The MODIS Fire Product [[12]] belongs to these global algorithms.

The algorithm uses brightness temperatures derived from the MODIS 4- and 11-μm channels, denoted by T_4 and T_{11}, respectively. The MODIS instrument has two 4-μm channels, numbered 21 and 22, both of which are used by the detection algorithm. Channel 21 saturates at nearly 500 K; and channel 22 saturates at 331 K. Since the low-saturation channel (22) is less noisy and has a smaller quantization error, T_4 is derived from this channel whenever possible. However, when channel 22 saturates or has missing data, it is replaced with the high saturation channel to derive T_4. T_{11} is computed from the 11-μm channel (channel 31), which saturates at approximately 400 K for the Terra MODIS and at 340 K for the Aqua MODIS. The 12-μm channel (channel 32) is used for cloud masking; brightness temperatures for this channel are denoted by T_{12}.

After the cloud- and water-covered pixels are identified, the potential fires are determined with the use of three conditions:

1) $T_4 > 310$ K; 2) $\Delta T = T_4 - T_{11} > 10$ K; and 3) $\rho_{0.86} < 0.3$,

where T_4 and T_{11} are brightness temperatures in channels 21/22 and 31 of the EOS/MODIS sensor; $\rho_{0.86}$ is the reflectance in channel 2 of this sensor.

Then, for background pixels, adjacent to the potential fires, the following statistical characteristics are determined: mean values (T_4*, T_{11}*, ΔT*) and mean absolute deviations (μ_4, μ_{11}, $\mu_{\Delta T}$) for T_4, T_{11}, and ΔT, respectively.

Further, the pixels, flagged as potential fires, are examined through the series of tests:

Test 1. $T_4 > 360$ K (320 K for nighttime pixels).
Test 2. $\Delta T > \Delta T^* + C_1 \cdot \mu_{\Delta T}$.
Test 3. $\Delta T > \Delta T^* + C_2$.
Test 4. $T_4 > T_4^* + C_3 \cdot \mu_4$.
Test 5. $T_{11} > T_{11}^* + \mu_{11} - C_4$.
($C_1 = 3.5$, $C_2 = 6$ K, $C_3 = 3.0$, $C_4 = 4$ K).

After testing, some pixel is classified as the fire, providing the following conditions are fulfilled:

a) test 1 or (test 2 + test 3 + test 4 + test 5) for daytime pixels;
b) test 1 or (test 2 + test 3 + test 4) for nighttime pixels.

To perform our comparative analysis, we took 1610 MOD14-type granules of the fire-dangerous season in 2003. These files contained results of detecting high-temperature anomalies, including geographical coordinates of "hot" pixels, satellite measurements at these points, and statistical characteristics in their vicinities. The official data of Tomsk Fire-Protection Service on forest fires in the territory of the Tomsk Region in June–September, 2003 were used as test data. The efficiency of the technology of early forest fire detection from space developed at the Institute (at least, on the regional level) is illustrated by table 4 presented below. It gives the data on the number of forest fires detected on the territory of the Tomsk Region, obtained by processing the AVHRR (IAO SB RAS) and MODIS (MOD14) satellite data.

Table 4. Results of comparison of the efficiency of satellite fire detection from the AVHRR (IAO) and MODIS (MOD14) data in the Tomsk Region in 2003. The number of early detected fires is given in the parentheses

	June	July	August	September	**Total**
AVHRR / IAO	16 (7)	60 (22)	82 (37)	28 (11)	186 (77)
MOD14	7 (4)	28 (11)	53 (16)	10 (6)	98 (37)
MOD14/ T	6 (3)	20 (6)	43 (13)	9 (6)	78 (28)
MOD14/ A	6 (4)	21 (7)	40 (8)	7 (4)	74 (23)

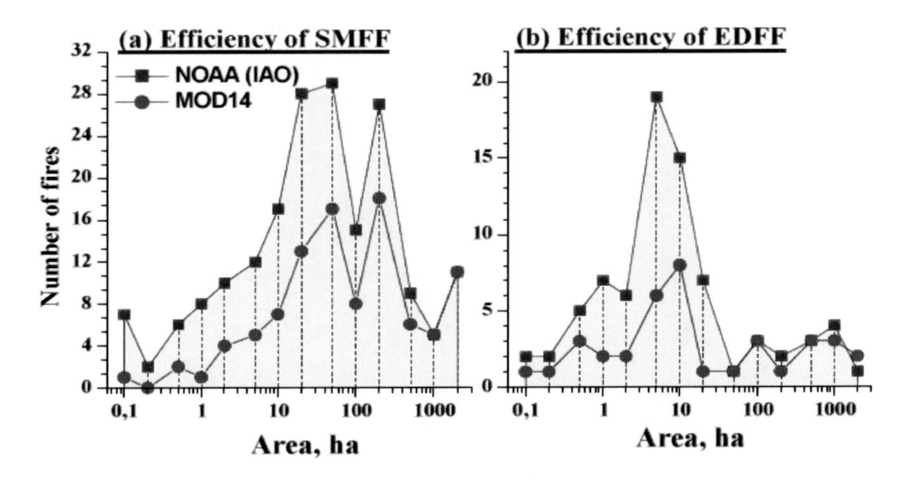

Figure 5. Results of comparative analysis of the efficiency of forest fire detection for the MOD14 and IAO algorithms.

Here we have used the following designations: MOD14 indicates that we used the MODIS data of both satellites (Terra and Aqua), MOD14/T indicates that we used only the data of the Terra satellite, and MOD14/A indicates that we used only the data of the Aqua satellite.

The data given in table 4 are supplemented with their graphic illustration (Fig. 5), where the estimates of the SMFF efficiency are given for both methods depending on the site area. The data presented in the Table above and in the figure give us grounds to conclude that the efficiency of the technology of early detection of forest fires from space developed at the IAO SB RAS is, at least on the regional level, a factor of two greater than that of the MODIS Fire Product algorithm used in the routine operation mode on the global scale.

Chapter 2

SOLUTION OF PROBLEMS OF THE TEMPERATURE MONITORING OF THE EARTH'S SURFACE FROM SPACE ON THE BASIS OF THE RTM METHOD

2.1. Formulation of the Problem of Fire Detection from Space

We now formulate basic relationships of the algorithm of reconstructing the brightness characteristics of a small-sized fire in the "surface + atmosphere + fire" system. Let a high-temperature object (fire) characterized by area S_F and temperature T_F ($T_F > 600$ K) be located on the surface of area S_{FOV} ($S_F << S_{FOV}$) corresponding to the field of view (FOV) of a remote sensor and having temperature T_S. The total upward thermal radiance at the top of the atmosphere I_λ (in the spectral range 3.5-4 μm) can be written as follows:

$$I_\lambda = B_\lambda(T_\lambda) = I_F + I_{BG}$$

$$I_F = \varepsilon_\lambda^F B_F P_\lambda,\ B_F = p(\theta)\, B_\lambda(T_F),\ p(\theta) = S_F / S_{FOV}(\theta)$$

$$P_\lambda = \exp\{-\tau_\lambda(\theta)\}$$

$$I_{BG} = I_{SRF} + I_{ATM} + I_{RFL} + I_{SCT}$$

$$I_{SRF} = (1-p)\, \varepsilon_\lambda^S B_\lambda(T_S)\, P_\lambda$$

where

$B_\lambda(T)$ is Planck's function

T_λ is the observed brightness temperature

I_F is the radiance of high-temperature object (fire) transmitted through the atmosphere

I_{BG} is the background radiance
P_λ is the atmospheric transmittance
τ_λ is the optical thickness of the atmosphere; $\tau_\lambda = \tau_\lambda^{mol} + \tau_\lambda^{aer}$ (molecular + aerosol)
θ is the viewing angle
I_{SRF} is the transmitted surface radiance
I_{ATM} is the atmospheric radiance
I_{RFL} is the ground reflected (thermal and solar) radiance
I_{SCT} is along-path scattered (thermal and solar) radiance
ε_λ^F is the fire emissivity ($\varepsilon_\lambda^F \approx 1.0$)
ε_λ^S is the background emissivity
T_S is the background temperature.

From the viewpoint of a correct consideration of optical-geometrical conditions of the observations, the problem of detecting the small-sized high-temperature object from space must be solved by reconstructing its radiance B_F:

$$B_F = (I_\lambda - I_{BG})/P_\lambda$$

where I_λ is the observed thermal radiance, and the values I_{BG} and P_λ are calculated from *a priori* optical-meteorological data. In this case, the decision rule for detecting forest fire from space $B_F > dB$ will be independent of the optical-geometrical conditions of observations.

Thus, the use of the RTM approach in the practice for detection of small-sized fires requires the fulfillment of the following key conditions:

- adequate model of IR radiative transfer through the atmosphere;
- real-time information: meteorological and optical parameters (of required volume and accuracy) of the atmosphere, geometry of the observations;
- “fast” software for atmospheric correction of the satellite IR images;
- retrieval of the land surface temperature (LST) in the channels λ=4 μm and λ=11 μm.

Next, a preliminary account of the following tasks is in order:

1) To obtain the statistical data on the variability of the values P_λ and I_{BG} for the NOAA/AVHRR infrared channels.
2) To estimate the relative contribution of I_{SRF}, I_{ATM}, I_{RFL}, and I_{SCT} to background radiance I_{BG}.
3) To demonstrate the dependence of I_{SCT} on the geometry of observations (θ, φ, Z).

To achieve the objectives, we simulated numerically the values P_λ, I_{SRF}, I_{ATM}, I_{RFL}, and I_{SCT} with the use of the LOWTRAN-7 computer code. In these computations, we used the actual meteorological parameters of the atmosphere and the geometry of satellite observations (θ, φ, Z) for Tomsk (56°30′ N, 85°00′ E) in May–September 1998-2000. More than 1300 situations were considered. The data were obtained from two satellites (NOAA–12 and NOAA–14) for two orbit types (antemeridian and postmeridian). The underlying surface was assumed Lambertian. The surface emissivities in the AVHRR IR channels 3 and 4 were $\varepsilon_{3.7}{}^{S} = 0.96$ and $\varepsilon_{11}{}^{S} = 0.98$. The near-surface air temperature was assumed to be equal to the underlying surface temperature T_S.

The data presented in Table 5 were recorded in channel 3 ($\lambda = 3.75$ μm) of the NOAA/AVHRR. Based on their analysis, we can conclude the following:

a. Total contribution of transmitted surface radiance I_{SRF} and atmospheric radiance I_{ATM} to I_{BG}, which linearly decrease with increasing aerosol optical thickness (AOT) τ_λ^{aer}, dominates.
b. Contribution of I_{RFL} is about 10% and decreases linearly with the AOT.
c. Contribution of I_{SCT} increases linearly with the AOT.
d. Even a significant growth of the AOT changes I_{BG} by 2-3% at most.

Table 5. Atmospheric transmittance and background radiance (λ= 3.75 μm) averaged over the period of observations (Tomsk, May–September 1999)

Visibility (*Vis*)	I_{BG}	P_λ	τ_λ^{aer}	Relative contribution to I_{BG} (%)		
				$I_{SRF}+I_{ATM}$	I_{RFL}	I_{SCT}
mol (no aerosol)	0.46435	0.74648	0	89.77	10.22	0.02
40 km, *rur*	0.46510	0.73267	0.01867	88.58	9.95	1.47
40 km, *urb*	0.46356	0.72971	0.02272	88.96	9.83	1.21
20 km, *rur*	0.46592	0.71791	0.03902	87.35	9.68	2.97
10 km, *rur*	0.46691	0.68934	0.07963	85.07	9.19	5.74
5 km, *rur*	0.46854	0.64510	0.14596	81.33	8.45	10.21
2 km, *rur*	0.47356	0.53298	0.33688	71.06	6.67	22.27
2 km, *urb*	0.45236	0.49339	0.41407	77.14	5.25	17.60
mol vs 2 km, *rur*	+1.98 %	- 28.60%				
mol vs 2 km, *urb*	- 2.58 %	- 33.90%				

Notes:

symbols *rur* and *urb* denote rural and urban boundary-layer aerosol models;

values of background radiance are expressed in $mW/(m^2 \cdot sr \cdot cm^{-1})$.

The temporal variability of the atmospheric transmittance and the background radiance in channel 3 (λ=3.75 μm) and channel 4 (λ=10.8 μm) of the NOAA/AVHRR can be estimated from the simulated data. We can conclude the following:

- The range of variations and the standard deviation of the atmospheric transmittance $P_{3.75}$ are three times less than those of $P_{10.8}$. This can be easily explained by a stronger dependence of $P_{10.8}$ on the atmospheric temperature and humidity, characterized by high spatiotemporal variability.
- On the other hand, the relative variability of background radiance I_{BG} in the channel with λ=3.75 μm is three times less than in the channel with λ = 10.8 μm. This is due to the fact that the relative temperature variations of Planck's function $B_\lambda(T)$ are a factor of 2.8 greater for λ=3.75 μm compared to λ=10.8 μm.

Because of the of complexity of computational algorithms, large volume of the required *a priori* information and difficulties in assigning actual *a priori* information with required accuracy, the calculation of I_{RFL} and I_{SCT} is the most labor- intensive problem. It is based on the knowledge of:

1. Geometrical conditions of observations (the view angle θ, solar zenith angle Z, and relative azimuth angle of measurements φ).
2. Meteorological parameters of the atmosphere.
3. Optical characteristics of the atmospheric aerosol.

In this regard, we consider some results of numerical simulation of the I_{SCT} value.

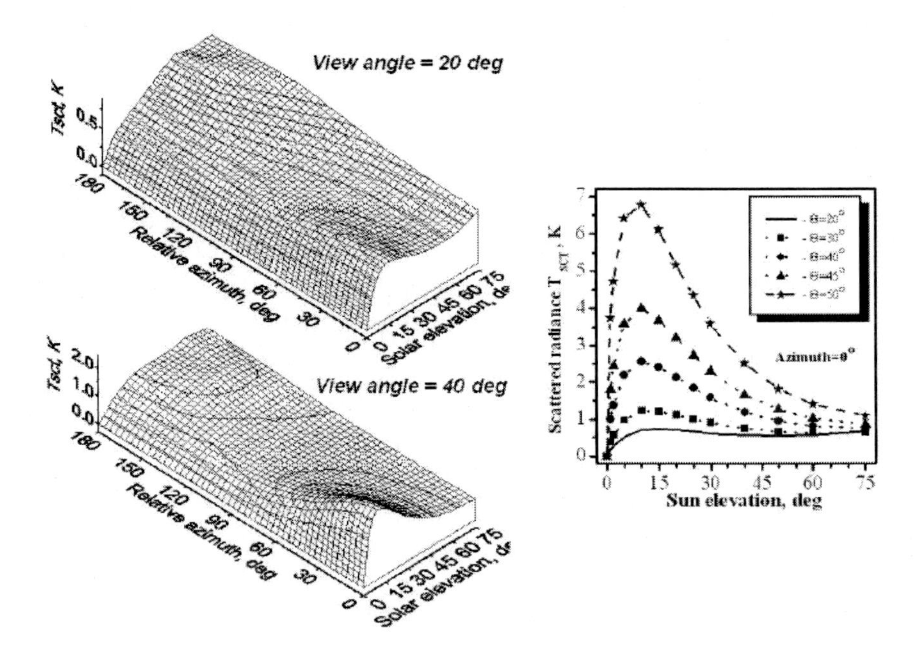

Figure 6. Dependence of the scattered radiance T_{SCT} on geometry of satellite observations; rural boundary-layer aerosol, *Vis* = 5 km.

To study the behavior of the scattered solar radiance in the channel with λ=3.75 μm, the I_{SCT} value was simulated numerically using the LOWTRAN-7 computer codes for a cloudless atmosphere and the following satellite observation conditions:

a. Meteorological parameters for the atmospheric model in the midlatitude summer.
b. Rural and urban boundary-layer aerosol models (visibility range *Vis* is from 50 to 2 km).
c. Geometry of observations: $\theta = 0 \div 55°$, $H_S = 90° - Z = 75 \div 0°$ (solar elevation angle), and $\varphi = 0 \div 180°$.

As can be seen from Fig. 6, the results of modeling demonstrate a complex dependence of the I_{SCT} value on the satellite observation conditions. For convenience, the I_{SCT} value in the figure is expressed as an increment T_{SCT} of the brightness temperature T_λ.

First of all, it should be noted that the azimuth dependence of T_{SCT} value becomes more pronounced with increasing scanning angle and decreasing *Vis* value (increase of the atmospheric turbidity). For solar elevation angle H_S of the order of 10° and azimuth angles $\varphi < 50°$, an anomalous local maximum of the T_{SCT} value is observed. The amplitude of this maximum depends on θ and on the optical characteristics of the boundary-layer aerosol.

An analysis of the NOAA satellite data on the relationship between the geometrical parameters H_S and φ for the Tomsk Region demonstrated the anomalous growth of T_{SCT}. This must be taken into account when correcting satellite measurements for the distorting effect of the atmosphere.

In this part, we study how the quality of the *a priori* meteorological information (AMI) affects the accuracy of reconstructing the object radiance from satellite data in the spectral range 3.5–4 μm. The atmospheric transmittance and the characteristics of upward radiation in the NOAA/AVHRR infrared channels were calculated for the atmospheric conditions in Tomsk with allowance for the actual geometry of satellite observations and meteorological parameters of the atmosphere specified in accordance with the IAO SB RAS data for the period of May–September 1998–2000.

Satellite measurements were simulated with the use of meteorological data closest in time to satellite observations. A high-temperature object with a temperature of 800–1200 K and area of 10–1000 m^2 was simulated within the radiometer field of view. Different types of AMI and sources of information about the background surface temperature T_S were used for atmospheric correction of the simulated "satellite measurements."

As a result, a correlation was found between the characteristics of AMI and the accuracy of radiance reconstruction for a high-temperature object. For an object with an area smaller than 100-200 m^2, the results demonstrate the marked effect of the AMI quality on the results of reconstruction. Depending on the AMI type, the RMS value of reconstructed radiance may vary several fold, but remains at least half as small as that without atmospheric correction.

2.2. Retrieval of Land Surface Temperature

In recent 25 years, of concern has been active development of satellite methods of LST retrieval [[17]-[24]], which generally called is "*split-window methods*" (SW methods). As a part of this approach, the IR measurements in two spectral channels of a "split" atmospheric transparency widow of 10–13 μm are used and the well-known method of differential absorption is implemented to account for the distorting water vapor effect.

Application of the SW-algorithm is based on linear relations between LST and satellite measurements in two spectral channels close to 11 and 12 μm. The relations can also include values of the underlying surface emissivity ε_λ for these channels. Parameters of these relations are calculated from modeling data or data of combined analysis of satellite and ground-based LST measurements.

As an example, we can consider the standard MODIS algorithm (MOD11) for remote LST measurement [[22]]:

$$T_0 = C + \alpha(T_{11} + T_{12})/2 + \beta(T_{11} - T_{12})/2;$$
$$\alpha = A_1 + A_2(1 - \varepsilon)/\varepsilon + A_3(\Delta\varepsilon/\varepsilon^2);$$
$$\beta = B_1 + B_2(1 - \varepsilon)/\varepsilon + B_3(\Delta\varepsilon/\varepsilon^2);$$
$$\varepsilon = (\varepsilon_{11} + \varepsilon_{12})/2,\ \Delta\varepsilon = (\varepsilon_{11} - \varepsilon_{12})/2,$$

where coefficients A_i and B_i ($i = 1, 3$) depend on the satellite zenith angle and integral moisture content of the atmosphere.

In practice, these algorithms are very simple and efficient for global LST monitoring. However, their users cannot disregard a number of serious practical limitations:

1. The LST retrieval error (δT_S) strongly depends on errors of δT_λ measurements. For instance, it is reported [[22]] that $\delta T_S \approx 6.19\delta T_\lambda$. For NOAA/AVHRR instrument, $\delta T_\lambda \approx 0.12$ K, i.e., $\delta T_S \approx 0.7$ K; while for EOS/MODIS system, $\delta T_\lambda \approx 0.05 \div 0.07$ K, i.e., $\delta T_S \approx 0.3 \div 0.4$ K.

2. The surface emissivities ε_{11} and ε_{12}, as well as their difference $\Delta\varepsilon$ should be well known. It is underlined in [[18],[19]] that at $\delta T_S \approx 0.5$ K the relative error $\delta\varepsilon$ of ε specification should be no more than about 0.5÷1%, and for $\Delta\varepsilon$ it is no worse than 0.25÷0.5%.

3. The coefficients of the algorithms are determined only for a given range of "standard" situations in the clear-sky atmosphere.

4. The algorithms take into account the thermal absorption by the water vapor; at the same time, the distortions, caused by aerosol and cirrus clouds, are disregarded.

Thus, the standard LST retrieval algorithms, used in practice, do not provide for confident and universal solution of the problem of atmospheric correction of IR measurements, especially under complex (non-standard) observational conditions.

Another, more correct approach involves the use of thermal radiative transfer models. The RTM method accounts for the distorting characteristics of the atmosphere with the use of widely known computer programs of the type of LOWTRAN, MODTRAN, ATCOR, etc., on the basis of the *a priori* optical-meteorological information on the atmospheric state at the moment of satellite observations. Examples of such an approach include the atmospheric correction of the MSU-SK, NOAA/AVHRR, Landsat, and ASTER [[25]-[28]] radiometer data.

Undoubtedly, this approach offers a universality and explicit accounting for all distorting factors in solution of the problem of LST retrieval from space, though its practical implementation requires invoking a large amount of real-time *a priori* information of the required quality and high-speed calculations. The intensive development of computation methods and modern technologies of parallel computer programming [[29]] eliminates labor consumptions of enormous computations. Moreover, a combined approach was suggested [[30]]: the fast SW method for standard situations and the RTM method for situations beyond the standard limits (in the presence of aerosol and semitransparent or cirrus clouds). The software package is also described in that same work, allowing a user, by means of accessible facilities (IMAPP and MODTRAN) and on the basis of EOS/MODIS satellite information, to employ the RTM method for the complex temperature monitoring of the Earth's surface, including LST retrieval and monitoring of high-temperature objects (HTO), i.e., fires and industrial thermal sources.

2.3. Distortions of Thermal Radiation by Molecular Atmosphere

It is well known the main factors of the molecular distortion of thermal radiation in EOS/MODIS channels include: the selective absorption by spectral lines of atmospheric gases and continuum absorption by line wings of H_2O and N_2. Though estimates of the influence of these factors on characteristics of upward fluxes of thermal radiation are available in the literature, the permanent development of thermal radiative transfer models necessitates some improvement of these estimates for a tighter relevance to the problems of the land surface temperature retrieval.

We used for these purposes the well-known software package LBLRTM_v11.3 (11/2007) [[31]], built upon the spectral line database HITRAN-2004 [[32]] (including all changes made by January 1, 2007) and molecular continuum models MT_CKD_2.1 [[33]].

2.3.1. Selective Absorption by Spectral Lines of Atmospheric Gases

The analysis of the HITRAN-2004 data on the total intensity of the molecular spectral lines and integrated gas content W_{GAS} allows one to separate out, from the total list consisting of 39 molecules, the optically active molecules (in the considered EOS/MODIS spectral channels), which determine the required accuracy of the LST retrieval by the RTM method: H_2O, CO_2, O_3, N_2O, and CH_4.

Figure 7 presents the absorption functions of thermal radiation, calculated with the use of LBLRTM_v11.3 in the considered EOS/MODIS channels. The influence of the selective absorption by each molecule (and their sum) on the accuracy of the RTM method can be estimated quantitatively by calculating the change of the radiation (brightness) temperature, measured in satellite channels, provided the chosen molecule is not taken into account in the line-by-line (*LBL*) calculations. Thus, it is necessary to determine the difference $\delta T_\lambda(\text{mol}) = T_\lambda(\sum) - T_\lambda(\sum\text{–mol})$, where $T_\lambda(\sum)$ and $T_\lambda(\sum\text{–mol})$ are calculated brightness temperatures, for which either all absorbing components ($\sum$) are taken into account or the chosen molecule ($\sum$–mol) is not taken into account.

Table 6 presents the δT_λ, estimates, allowing us to draw certain conclusions:

1) First of all, obviously, the influence of the selective absorption by atmospheric gases in all EOS/MODIS channels exceeds the 0.25 K level and, hence, should be accounted for within the RTM method.

2) In channels 20 and 21, the distorting effect of the selective absorption is determined by lines of H_2O, N_2O, and CH_4 molecules.

3) In channels 31 and 32, it is sufficient to take into account only the contribution of H_2O lines, with much less accounting for the CO_2 line contribution.

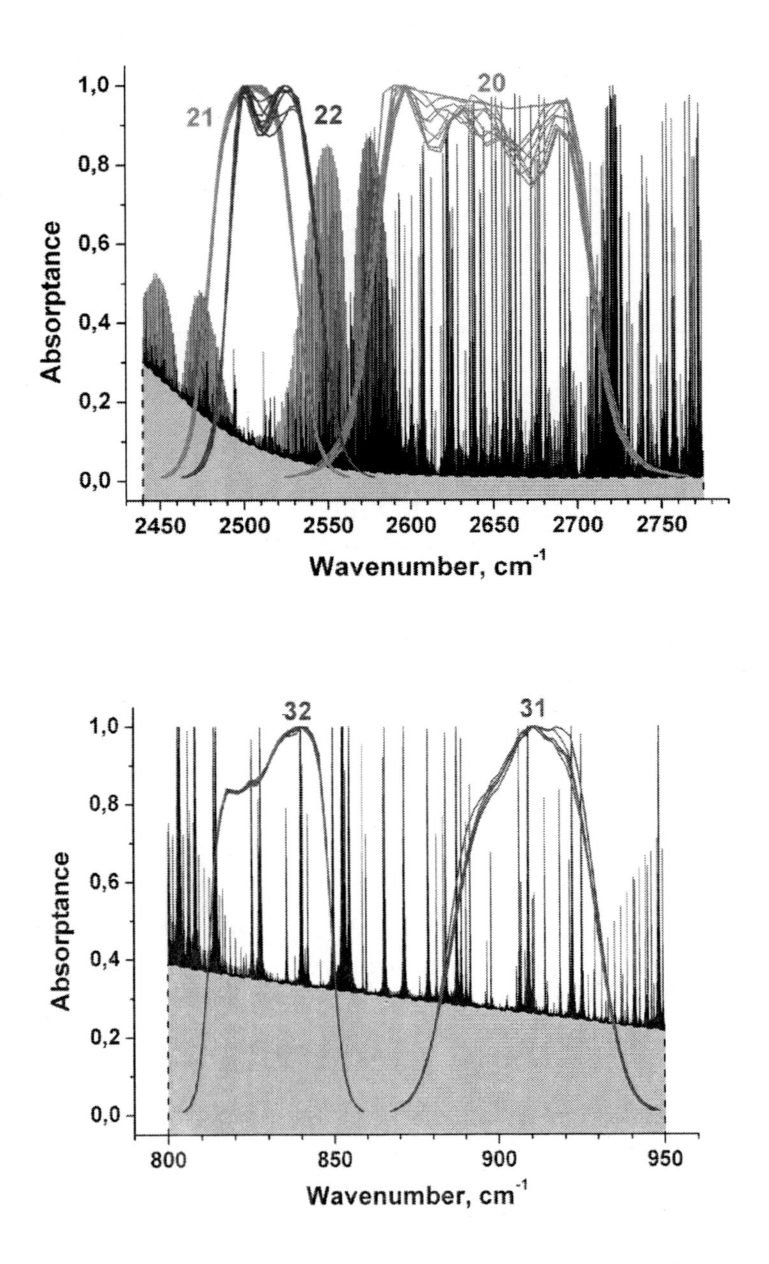

Figure 7. Absorption of thermal radiation in 20, 21/22, 31, and 32 EOS/MODIS spectral channels. Midlatitude summer. Shade of grey shows continuum; peaks are for lines + continuum.

Table 6. Optical depth τ of atmospheric gases and distortion of brightness temperature δT_λ(mol), K. *Midlatitude summer*

Molecules	Spectral channel							
	20		21		31		32	
	τ	δT_λ	τ	δT_λ	τ	δT_λ	τ	δT_λ
LBLRTM_v11.3 data								
H_2O	0.1267	0.935	0.0035	0.020	0.0859	0.685	0.0827	0.662
CO_2	0.0013	0.022	0.0017	0.027	0.0027	0.034	0.0049	0.070
O_3							0.0003	0.015
N_2O	0.0175	0.319	0.0174	0.260				
CH_4	0.0102	0.145	0.0045	0.067				
Other molecules							0.0007	0.015
All molecules	0.1572	1.420	0.0273	0.376	0.0915	0.857	0.0889	0.765
All (tropics)	0.1949	1.923	0.0298	0.436	0.1209	1.151	0.1155	0.994
MODTRAN_v3.5 data								
H_2O	0.1288	0.926	0.0040	0.022	0.0756	0.615	0.0911	0.772
All molecules	0.1584	1.416	0.0315	0.429	0.0876	0.969	0.0979	0.896
All (tropics)	0.1969	1.917	0.0342	0.496	0.1126	1.231	0.1239	1.119

Thus, in the framework of the RTM method, the problem of fast specification of the confident *a priori* information concerns only the temperature and humidity profiles.

One more important condition of the successful use of the RTM method in practice is a good accuracy and high speed of calculation of selective absorption coefficients in the processing of large-volume satellite information. Obviously, the direct use of the *LBL* methods in the framework of the RTM method is impossible in view of their laboriousness; therefore, it is advisable to use simplified radiative transfer models, tested in practice and accessible to wide user community, such as the commonly known MODTRAN program. Presently, the program MODTRAN_v4.x [[35]] is a commercially available product; however, its predecessor MODTRAN_v3.x [[34]] and its codes are accessible (that is important) to users. Table 6 presents δT_λ(mol) calculations with the use of MODTRAN_v3.5 program, based on parameters of spectral lines from HITRAN-96 [[36]] and models of molecular continuum CKD_v2.1_rev.3.3 [[37]]. The comparison of δT_λ(mol) values, obtained by MODTRAN_v3.5 and LBLRTM_v11.3, suggests that these data disagree by less than 0.15 K, reasonably well meeting the practical accuracy requirements for the satellite-borne LST retrieval.

2.3.2. Continuum Absorption by Spectral Line Wings of Atmospheric Gases

According to the MT_CKD_v2.1 model, in addition to the selective absorption of the thermal radiation by lines, located inside the spectral channels, a marked influence is exerted by the continuum absorption by line wings of intense H_2O, CO_2, O_3, and N_2 bands, lying outside these spectral channels (see the figure 7). The atmospheric transparency window 3.5–4 μm is characterized by a weak H_2O continuum, while at wavenumbers $\nu < 2600$ cm^{-1} there is a stronger N_2 continuum. The transparency window of 10–13 μm is dominated by strong H_2O continuum absorption, with a simultaneous presence of a weak CO_2 continuum. The H_2O continuum resolves into two components, corresponding to self-broadening of lines (H_2O–H_2O) and air-caused line broadening (H_2O–AIR).

To quantitatively estimate the influence of each continuum component, we calculated δT_λ(cont) in analogy to δT_λ(mol) calculations: $\delta T_\lambda(\text{cont}) = T_\lambda(\sum) - T_\lambda(\sum - \text{cont})$, where $T_\lambda(\sum)$ and $T_\lambda(\sum - \text{cont})$ are the calculated brightness temperatures, for which all absorbing components ($\sum$) are taken into account or the chosen continuum component ($\sum$–cont) is not accounted for.

Table 7 presents the δT_λ(cont) calculations, whose analysis allows us to make the following conclusions.

1) The effect of H_2O and CO_2 continuums on the brightness temperature in channels 20 and 21 is less than 0.05 K. The effect of N_2 continuum in channel 20 has the same order of magnitude; however, it markedly increases in channel 21, exceeding a level of 1 K.

2) The component H_2O–H_2O ($\approx$ 1–2 K) dominates in channels 31 and 32, with a much weaker (less than 0.2 K) influence of the component H_2O–AIR. The effect of CO_2 continuum is almost insignificant (less than 0.01 K).

3) Comparing the δT_λ(cont) values, obtained via LBLRTM_v11.3 and MODTRAN_v3.5 programs, we see that they disagree in channels 20, 31, and 32 by less than 0.1 K, and they increase up to 0.2 K due to N_2 continuum only in channel 21.

Table 7. Optical depth τ of molecular continuum and distortion of brightness temperature δT_λ(cont), K. Midlatitude summer

Components	Spectral channel							
	20		21		31		32	
	τ	δT_λ	τ	δT_λ	τ	δT_λ	τ	δT_λ
LBLRTM_v11.3 data								
H_2O–H_2O	0.0019	0.008	0.0029	0.013	0.2959	1.400	0.3956	1.825
H_2O–AIR	0.0026	0.014	0.0001	0.000	0.0112	0.072	0.0278	0.174
CO_2	0.0003	0.003	0.0034	0.041	0.0001	0.002	0.0003	0.006
N_2	0.0058	0.067	0.1052	1.251				
All	0.0106	0.093	0.1115	1.309	0.3072	1.483	0.4237	2.032
All (tropics)	0.0131	0.119	0.1135	1.454	0.5525	3.064	0.7558	4.060
MODTRAN_v3.5 data								
H_2O–H_2O	0.0044	0.020	0.0061	0.027	0.3188	1.491	0.4390	1.986
H_2O–AIR	0.0034	0.018	0.0002	0.001	0.0008	0.005	0.0049	0.030
N_2	0.0074	0.093	0.1174	1.445				
All	0.0152	0.131	0.1236	1.475	0.3196	1.497	0.4439	2.022
All (tropics)	0.0199	0.173	0.1278	1.647	0.5849	3.170	0.8099	4.144

2.3.3. The Influence of Errors in Setting Profiles of Meteorological Parameters

To date, the current databases of spectral line parameters, molecular continuum models, and thermal radiative transfer models, overall, ensure a high accuracy of accounting for the distorting influence of the atmosphere, when using the a priori valid information on key meteorological parameters of the atmosphere X(z), where z is the height. Since the vertical profiles of X(z) contain measurement (retrieval) errors δX(z), it seems reasonable to estimate the effect of these errors on the accuracy of the RTM method.

The estimates were made as follows: 1) for the chosen profile of atmospheric meteorological parameters (e.g., meteorological model of the midlatitude summer), we calculated the brightness temperature $T_\lambda(0)$; 2) some changes δX(z) were introduced in a given profile and the value of $T_\lambda(\delta X)$ was calculated for the distorted profile; and 3) the difference $\delta T_\lambda(\delta X) = T_\lambda(0)$ –

$T_\lambda(\delta X)$ was calculated, being the measure of the influence of the inaccuracy in setting meteorological parameters on the brightness temperature.

Table 8. Change of brightness temperature caused by variations of the profiles of meteorological parameters: the air temperature δT_{AIR}, the humidity δW_{H_2O}, and the minor atmospheric gas content δW_{GAS}. LBLRTM_v11.3 data

Parameter	Spectral channel			
	20	21	31	32
Midlatitude summer				
$\delta T_{AIR} = +2$ K	+0.206	+0.150	+0.632	+0.786
$\delta W_{H_2O} = +20\%$	–0.153	–0.010	–0.659	–0.820
$\delta W_{GAS} = +40\%$	–0.168	–0.151	–0.068	–0.043
Tropics				
$\delta T_{AIR} = +2$ K	+0.241	+0.147	+0.968	+1.170
$\delta W_{H_2O} = +20\%$	–0.218	–0.020	–1.199	–1.418
$\delta W_{GAS} = +40\%$	–0.186	–0.169	–0.075	–0.043

Table 8 presents the calculations of $\delta T_\lambda(\delta X)$ for the air temperature and moisture content, as well as the content of other atmospheric gases. For the air temperature and moisture content we have chosen $\delta T_{AIR} = +2$ K and $\delta W_{H_2O} = +20\%$ at all atmospheric levels. They can be considered as characteristic retrieval errors of atmospheric meteorological parameters according to EOS/MODIS data [[38]]. For profiles of other atmospheric gases, we have chosen $\delta W_{GAS}=+40\%$ as a certain limiting value. Thus, the data of Table 8, overall, reflect the maximal effect of meteorological parameter profile errors on T_λ. Considering for the limiting character of these estimates, we can make the following conclusions.

1) In channels 20 and 21, the influence of variations of profiles of all meteorological parameters in absolute value is less than 0.25 K, which, in principle, permits one in practice to optimize the volume of calculations of the distorting atmospheric parameters.

2) In channels 31 and 32, the value of δT_λ for a given δW_{GAS} does not exceed 0.1 K; therefore, the setting of *a priori* information on the atmospheric content of minor gas constituents in these channels does not require a high accuracy. At the same time, the effect of uncertain setting of the temperature and air humidity profiles is significant ($\delta T_\lambda > 0.5$ K) for the correct treatment

of the molecular distortion of the thermal radiation in the framework of the RTM method. Note that the absolute value of δT_λ is determined by the degree of the thermal radiation absorption in the channel; therefore, $|\delta T_\lambda|$ is less in channel 31 than in channel 32. This circumstance can be used to compensate for the effect of imperfect setting of meteorological parameters in the RTM method.

3) Note that identical signs of δT_{AIR} and δW_{H_2O} correspond to differently signed δT_λ values. That is, in the presence of the positive correlation between δT_{AIR} and δW_{H_2O}, this circumstance may lead to mutual error compensation in setting meteorological parameters, which are key ones to the atmospheric correction. Thus, the atmospheric correction of remote IR measurements of LST becomes possible on the basis of the meteorological information with relatively low accuracy characteristics.

The latter two conclusions should be complemented with some important comments. First, the analysis of the satellite methods of retrieval of the temperature and humidity profiles [[38]] allows us to suppose with a high degree of confidence that errors in the retrieval of the temperature and humidity have just a positive correlation. Second, the difference between $\delta T_\lambda(\delta X)$ values in channels 31 and 32 allows us to propose the compensation for the δX effect through application of the RTM method by the "*split-window*" principle, that is, the LST is to be determined as follows: $T_S = T_{S,31} - \Delta T_S$; $\Delta T_S = C_{ERR}(T_{S,32} - T_{S,31})$, where $T_{S,31}$ and $T_{S,32}$ are LST values, retrieved in 31 and 32 channels; $C_{ERR} \approx 2.0$ is the coefficient obtained from simulation calculations. This will ensure the RTM method resistance to uncertainties in setting the *a priori* meteorological information.

Chapter 3

SOFTWARE AND VALIDATION OF MODIS ATMOSPHERIC DATA

3.1. STRUCTURE OF THE PROGRAM COMPLEX

Note firstly that two types of the basic software for thematic processing of information from EOS/MODIS system are known to date.

The first type includes DRL licensing programs (Direct Readout Lab, GSFC/NASA), where basic algorithms are grouped in PGE (Product Generation Executive), which include the program texts and necessary data for their assembling, setting, and exploiting. In Russia, this software is successively used for many years at the Center of Space Monitoring of the Altay State University.

The second type is represented by the well-known program package IMAPP (International MODIS/ AIRS Processing Package). The design, maintenance, and distribution of the Package are conducted under GNU General Public License at the Space Science and Engineering Center (SSEC), being a division of the University of Wisconsin–Madison (ftp://ftp.ssec.wisc.edu/pub /IMAPP/MODIS/).

To solve our problem, we chose the IMAPP v. 2.0 package and adapted it to the operation medium Windows. General scheme (Fig. 8) of the designed software for thematic processing of the EOS/MODIS data includes three stages.

1) At the initial stage (levels 0 and 1), the satellite file EOS/MODIS is unpacked with the help of IMAPP program from PDS format to a set of HDF-

EOS formats; the geographic assignment of data and calibration of the space-borne measurements are performed.

2) At the second stage (level 2) the *a priori* information from MODIS on parameters of the atmospheric state is prepared for processing. This information includes:

– cloud mask (MOD35);

– optical characteristics of aerosol (MOD04);

– total column precipitable water vapor (MOD05);

– characteristics of clouds (MOD06);

– atmospheric profiles of the geopotential height, temperature, and humidity of air, ozone content (MOD07).

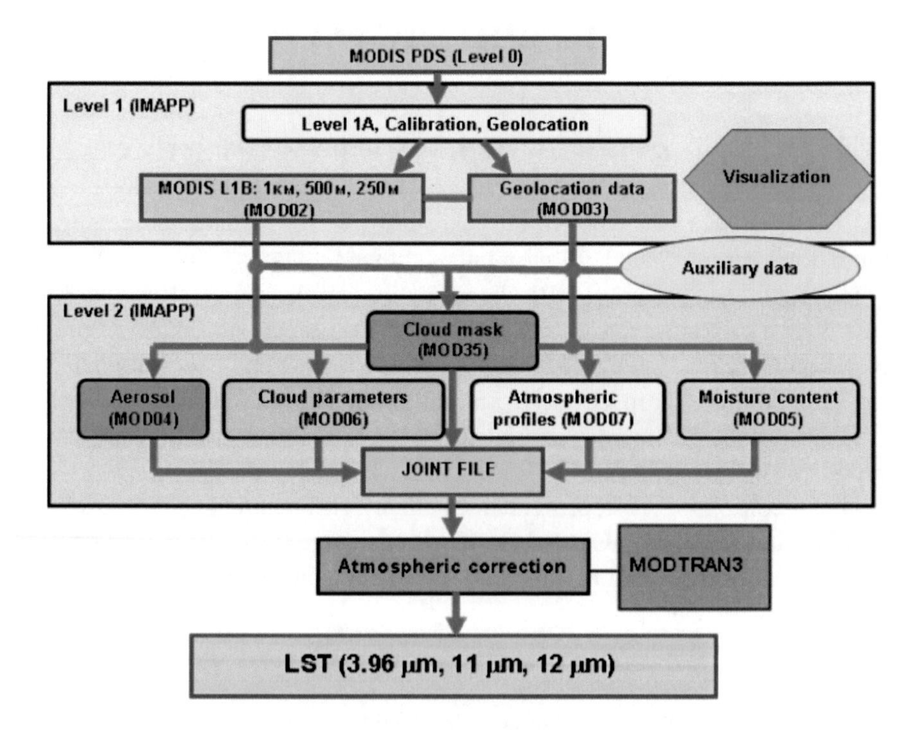

Figure 8. Schematic view of the atmospheric correction of remote measurements of the LST with the use of the EOS/MODIS satellite system.

3) Based on the *a priori* information, formed at the stage 2, with the use of the program block "Atmospheric correction", characteristics of distorting effects of the atmosphere are calculated and the space-borne measurements of

LST and its spectral reflectance are corrected. At present, this is made by the well-known program MODTRAN.

To set values of the emissivity ε_λ in the IR channels at λ=3.96, 11, and 12 μm of EOS/MODIS device, data of MODIS UCSB Emissivity Library (http: // www.icess.ucsb.edu / modis / EMIS / html / em.html) or the database Global Infrared Land Surface Emissivity Database (http://cimss.ssec.wisc. edu/ iremis/) are used.

3.2. Validation of MOD07 Data (Atmospheric Profiles)

One of the main stages of the real-time atmospheric correction of satellite IR measurements involves accounting for the molecular absorption on the basis of the atmospheric profiles (air temperature and humidity), retrieved from MODIS data.

Two main approaches, namely, the physical and regression methods, have been developed for this. The first ("exact") method is based on the iterative solution (with the use of regularization) of the system of nonlinear radiative transfer equations in the water vapor and carbon dioxide IR absorption bands. However, in view of the complexity and computation overburden of this method, in practice the large volumes of MODIS satellite data are processed using "fast" statistical method, where the solution is based on the system of linear regression equations.

The literature contains publications (see, e.g., [[38],[39]] which, in addition to the physical and mathematical foundations of these methods, present certain data on their accuracy. Unfortunately, these data cannot answer correctly the question about how well does the accuracy of these methods correspond to the requirements of the physical (RTM) approach to the LST retrieval with an error less than 0.5-1 K. To study this question, it is necessary to compare the atmospheric profiles, obtained from satellite-based and test (e.g., radiosonde) measurements. Also, it is expedient to compare the results of the calculations, where the atmospheric profiles according to MODIS or radiosonde data are used to calculate the brightness temperatures T_λ and to retrieve LST from satellite IR measurements.

3.2.1. Initial Data for Validation

First of all, we should make a remark concerning the selection of the test data for validation of the satellite measurements of meteorological parameters. Evidently, test profiles, obtained synchronously with satellite measurements, would be most correct. However, such subsatellite experiments require quite serious financial expenses; therefore, in practice the satellite data are usually validated with the help of standard air temperature and humidity measurements, which are performed regularly at radiosonde stations.

The main problem, posed in this work, had been solved in two stages. At the first stage, we performed a comparative analysis of results of retrieval of atmospheric profiles from data of EOS/MODIS satellite system and results of the test radiosonde measurements. At the second stage, the obtained test sample of radiosonde and satellite data, collocated in time and space, was used to calculate, with the help of the MODTRAN_v.3.5 program, the brightness temperatures T_λ, and to imitate the LST retrieval in the IR channels (31 and 32) of MODIS.

For the studies, we used the following initial data:

- results of radiosonde measurements of the vertical profiles of air temperature and humidity for May-September 2003 at 115 aerological stations in Russia (NOAA/ESRL/GSD – RAOB website http://raob.fsl.noaa.gov/), times of measurements are 00:00 and 12:00 GMT;
- files with vertical profiles of the atmospheric meteorological parameters (air temperature and humidity), retrieved from MODIS measurements onboard Terra and Aqua satellites (LAADS Web website http://ladsweb.nascom.nasa.gov/data/search.html).

Obviously, the radiosonde and satellite data are compared under an unavoidable allowance that the local and spatial data are not coincident in space and time. For instance, in work [[39]] the ground-based and satellite measurements were admitted to be separated by up to two hours, and satellite data were allowed to be not collocated with the point of radiosonde measurements to within $1\times1°$ latitude by longitude. To increase the correctness of the validation results, the satellite data were selected such that the maximum time lag between the satellite and ground-based measurements did not exceed 90 minutes, and both were within 25 km of each other. Moreover, the test sample was prepared to include only those satellite data which, according to cloud mask, contained no more than 30% of cloud per a pixel. In addition, we sorted out the possible situations of the meteorological measurements under conditions of snowfall, dense cirrus clouds, and rain showers.

We separate the latitude belt 45-62°N and classify the stations by the territorial attribute into 3 groups: West (30-60°E), Siberia (60-90°E), and East (90-180°E). The region of the West Siberia includes Omsk, Tomsk, Novosobirsk, Kemerovo regions and Altai Krai. We can conclude that the test data for the West Siberia will have minimum information content to perform the validation. Indeed, the test sample for this region contains only 12 pairs of profiles, obtained at the meteorological station in Omsk and according to measurements onboard Aqua satellite. At the same time, the bulk of the test information is the data of radiosonde stations, located in the West (about 2000 profiles) and the East (of the order of 1200 profiles).

For the test sample, the near-ground air temperature was in the wide range from -7 °C to 34°C, and the total moisture content varied from 0.3 to 4.5 cm. It should also be noted that, despite the fact that all stations worldwide conduct everyday radiosonde measurements only at t=00:00 and 12:00 GMT, the data of the test sample pertain to a wide diurnal range of local times. That is, work dealing with validation of MODIS satellite meteorological data is based on the test aerological information which is characterized by wide spatiotemporal variability range of meteorological parameters.

3.2.2. Analysis of Validation Results

The satellite and aerological measurements of the weighted average temperature and total moisture content of the atmosphere are compared in Figure 9 and Table 9.

It contains the average discrepancies (μ), their standard deviations (σ), as well as the correlation coefficients (R). Entries are categorized by the type of satellite (Terra and Aqua) and by the time of measurements (t=00:00 GMT and 12:00 GMT).

First of all, we should point out quite a high correlation coefficient between the aerological and satellite measurements: R=0.90-0.99 for temperature and R=0.81-0.90 for moisture content. The discrepancies δT and δW between the test and satellite data have one and the same sign for all the four groups, and the values $10 \cdot \delta T$ and δW are close in the order of magnitude. According to data of Table 9, this result allows us to legitimately speculate that the calculation of the radiation temperatures $T_{\lambda,R}$ and $T_{\lambda,M}$ with the use of radiosonde or satellite measurements of the atmospheric profiles shall give close results.

To verify this speculation, the MODTRAN_v.3.5 program was used to calculate $T_{\lambda,R}$ and $T_{\lambda,M}$ in the MODIS channels 31 and 32. Table 10 presents the results of these calculations. For entire sample, the discrepancies δT_λ between $T_{\lambda,R}$ and $T_{\lambda,M}$ for the channel 31 are $\mu(\delta T_\lambda)$=0.17 K and $\sigma(\delta T_\lambda)$=0.46 K, and the correlation coefficient between $T_{\lambda,R}$ and $T_{\lambda,M}$ does not exceed the level R=0.99. For the channel 32, the analogous values are $\mu(\delta T_\lambda)$=0.23 K and $\sigma(\delta T_\lambda)$=0.62 K.

A close result was obtained when LST retrieval was imitated on the basis of RTM method with the use of the atmospheric profiles according to MODIS data. In this case, the average LST retrieval error in the channel 31 was $\mu(\delta T_S)$=-0.22 K, and $\sigma(\delta T_S)$ did not exceed 0.60 K. For readily explainable reasons, in the channel 32 we obtained larger errors $\mu(\delta T_S)$=-0.29 K and $\sigma(\delta T_S)$=0.88 K.

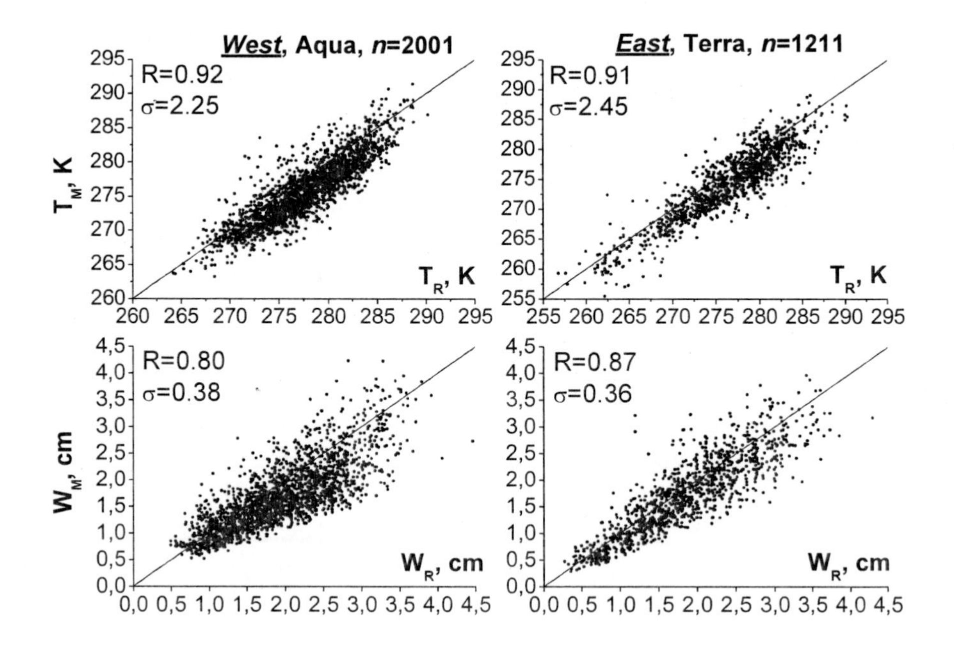

Figure 9. Results of comparison of MODIS (M) and radiosonde (R) data on the weighted mean temperature (T) and moisture content (W) of the atmosphere. Superimposed on the plots is the line y=x.

Table 9. The statistical characteristics (μ σ R) of the discrepancies δT and δW in the profiles of the atmospheric temperature and moisture content, obtained according to radiosonde data and onboard satellites (Terra and Aqua); *n* is the sample size, semicolon separates the μ and σ values

Region	Terra		Aqua	
	δT, K	δW, %	δT, K	δW, %
Time t=00:00 GMT				
West n=1595	–		2.81; 1.87 R=0.910	28.3; 26.6 R=0.808
East n=399	0.70; 2.60 R=0.904	5.5; 20.6 R=0.891	0.51; 2.21 R=0.934	11.8; 20.4 R=0.900
Time t=12:00 GMT				
West n=417	-0.11; 1.69 R=0.925	-17.4; 12.2 R=0.939	0.19; 2.57 R=0.930	0.3; 19.7 R=0.853
East *n*=799	3.26; 2.06 R=0.940	19.9; 26.0 R=0.874	—	

Table 10. The statistical characteristics (μ σ) of the discrepancies in the T_λ values calculated on the basis of radiosonde and satellite data or retrieved LST; semicolon separates the μ and σ values

Region	Terra		Aqua	
	Channel 31	Channel 32	Channel 31	Channel 32
Time t=00:00 GMT				
West	–		0.15; 0.39 -0.19; 0.49	0.19; 0.53 -0.24; 0.73
East	0.10; 0.45 -0.13; 0.59	0.15; 0.61 -0.19; 0.89	-0.04; 0.39 0.04; 0.47	-0.05; 0.53 0.07; 0.70
Time t=12:00 GMT				
West	0.29; 0.25 -0.36; 0.32	0.40; 0.33 -0.52; 0.48	0.03; 0.61 -0.08; 0.83	0.04; 0.81 -0.10; 1.26
East	0.37; 0.50 -0.47; 0.64	0.51; 0.67 -0.67; 0.97		
Total	0.29; 0.54 -0.36; 0.70	0.39; 0.72 -0.51; 1.05	0.12; 0.41 -0.15; 0.52	0.15; 0.55 -0.19; 0.76

Items in the first line are results of T_λ simulation.

Items in the second line are results of LST retrieval.

We should also note the following. The simulations gave fairly small error of LST retrieval with the use of RTM method; nevertheless, here we also had a small number (less than 2%) of situations with substantial discrepancies between the test and satellite meteorological data, which had led to LST retrieval errors as large as several degrees.

To solve this problem, it is reasonable to use the approach, which we suggested in work [[41]]; specifically, the influence of errors in specifying the profiles of the meteorological parameters during atmospheric correction of the satellite LST measurements is mitigated by applying the "*split-window*" principle to LST values, retrieved with the help of the RTM method simultaneously in the MODIS channels 31 and 32 (see section 2.3.3). By applying this relationship to the calculated $T_{S,31}$ and $T_{S,32}$ values (obtained in this work according to the numerical simulation data), it was shown that the LST retrieval error in this case had decreased to the level less than 0.1 K. These results are also depicted in Fig. 10. Noteworthy, the maximum LST retrieval errors decreased to 1.2 K, and δT_S values exceeded the level of 1 K only in 5 cases.

In conclusion of the work, the data of simulation of the radiation temperatures $T_{\lambda,M}$ were used to test the regression SW algorithm [[21],[22]] of retrieval of the surface temperature. In this case, the error of the SW algorithm was: $\mu(\delta T_{SW})$=0.39 K and $\sigma(\delta T_{SW})$=0.12 K; and the number of the situations with $\delta T_{SW}>1$ K now did not exceed 1.1% (866 cases). Thus, during treatment of the molecular absorption, the standard deviation for the regression algorithm turned out to be lower than that from the use of one-channel RTM method. However, the "two-channel" RTM method already demonstrates higher accuracy characteristics compared to the regression algorithm.

The authors express gratitude to the officials and personnel of NASA, whose effort was invaluable in obtaining the MOD07_L2 satellite data used in this paper. Authors also thank the officials and staff of NOAA/ESRL/GSD for providing the possibility to use in the work the data from the RAOB archive, containing the radiosonde measurements of the vertical profiles of the atmospheric temperature and humidity.

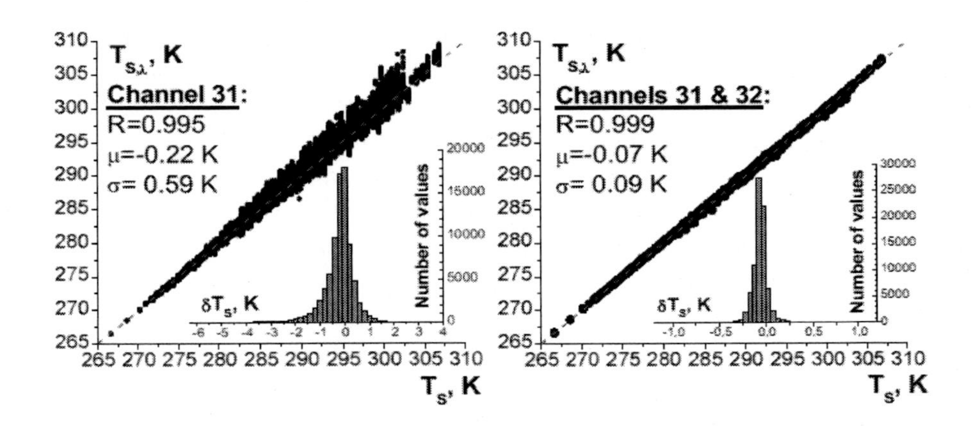

Figure 10. Results of LST retrieval on the basis of RTM approach with the use of MODIS satellite meteorological data: a) channel 31 (λ=11 μm) and b) combination of the channels 31 and 32. Insets give histograms of the frequency distribution of LST retrieval errors.

3.3. Validation of MOD04 Data (Aerosol Optical Depth)

After survey of scientific publications, available in Internet, we can state that for the territory of Siberia there are no fully correct data on the accuracy of the algorithm [[43]] of aerosol optical depth (AOD) measurements from MODIS satellite. In this regard, a complex study of this issue seems to be justified.

Comparison with AERONET Data

In accordance with the method [[44]], a regional validation of satellite AOD measurements begins with a creation of test sample, which includes the satellite and ground-based AOD measurements after their time and space collocation. To this end, for a given AERONET site and a chosen time period, we select the files (granules) of MOD04_L2 satellite data, for which the geographic coordinates of the station are within the image. Next, we determine the position of the pixel nearest to the location of the station. It is the center of 5×5 pixel rectangular test "window", whose nominal size is 50×50 km in this case. For this window, the statistical characteristics of AOD values in the spectral channels are calculated. Then, the ground-based photometric measurements within ±30 min of the satellite measurements are selected, and for them the statistical characteristics are also calculated. The obtained test sample of ground-based and satellite data is then subject to statistical analysis, i.e., the average discrepancies (μ) of the data, their standard deviations (σ), and the correlation coefficients (R) are determined. It is important to keep in mind the differences in the spectral characteristics of the measurement channels of photometer (λ=440 nm, 500 nm, and 675 nm) and MODIS instrument (λ=466 nm, 550 nm, and 646 nm).

In our case, the test sample was created on the basis of data obtained in the period spring-fall of 2001-2009 onboard Terra and Aqua satellites and during ground-based photometric measurements at AERONET stations in Tomsk (since 2002), Krasnoyarsk (2001-2004), and Irkutsk (since 2005).

For the Terra satellite, Figure 11 presents the results of validation of satellite AOD measurements (numbering n=397) in the spectral channel λ=550 nm at the AERONET station in Tomsk. Recall that the $\tau_{0.55}$ value, together with the meteorological visibility range, is used as an input parameter for specifying the aerosol characteristics in the MODTRAN program. Data of ground-based photometric measurements of AOD for this wavelength were obtained via interpolation according to the Angstrom formula $\tau_\lambda=\beta\lambda^{-\alpha}$. Also shown in the figure is the dependence of the validation results on the state of clouds in the test window, characterized by the relative number N_{CLD} of "cloudy" pixels (a maximum of 96%, or one clear-sky pixel) in the test window.

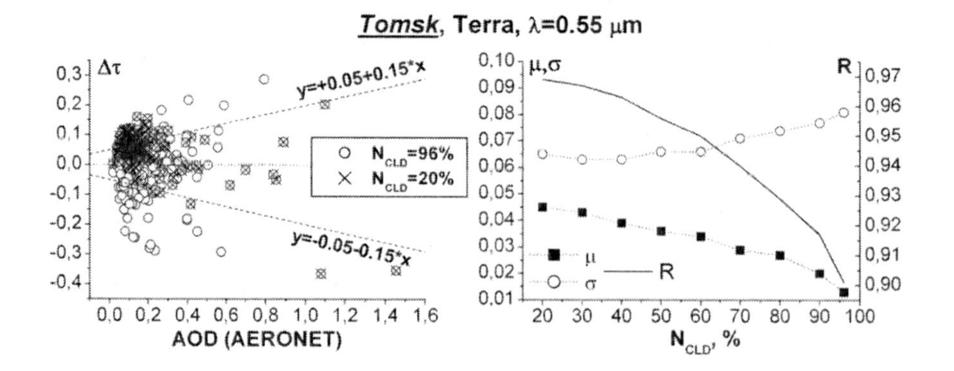

Figure 11. Results of validation of satellite AOD measurements; statistical data (μ,σ,R) of discrepancies of the ground-based and satellite AOD measurements for different cloud conditions.

Data of Fig. 11 suggest that, with increase of N_{CLD}, the correlation coefficient R between the ground-based and satellite AOD measurements decreases (from 0.97 to 0.90), and, at the same time, dispersion of the discrepancies increases from σ=0.065 to σ=0.080. Surprisingly, as N_{CLD} grows, the average discrepancy between the ground-based and satellite measurements markedly decreases, from μ=0.050 to μ=0.013. Analysis of data for AERONET stations in Krasnoyarsk and Irkutsk or Aqua satellite gives similar results.

Results of validation of satellite AOD measurements, obtained for Siberian region, were complemented with data of integrated aerosol experiment, performed in spring 2009-2010 in Primorye and the Sea of Japan [[45]]. In the frameworks of the experiment, we performed a comparative analysis of satellite AOD measurements and ground-based photometric measurements at the AERONET stations in Ussuriysk, China, South Korea, and Japan. The validation results, obtained in this experiment, confirmed the main conclusions, made for the AERONET stations in Tomsk, Krasnoyarsk, and Irkutsk.

Spatial Analysis of Data

In addition to the estimate of the accuracy of "local" satellite AOD measurements at locations of AERONET stations, our works [[46],[47]] have suggested and performed the spatiotemporal analysis of MOD04 data as an extra estimate of their accuracy. As an example of analogous studies, we should present the results of work [[48]], where the spatial analysis of satellite AOD measurements was performed for the region of Mediterranean Sea.

The seasonally averaged spatial distributions $\tau(x,y)$ were analyzed for the region 55–62°N, 74–90°E. The spatial averaging was performed for the surface regions with the size of 0.5° latitude by 1° longitude, their nominal area being of the order of 50×50 km. The results of $\tau(x,y)$ analysis, presented in Fig. 12, showed that the spatial AOD fields are inhomogeneous. The structure of these inhomogeneities is similar to the spatial distribution of the types of the underlying surface in the region. In the absence of local intense sources of aerosol pollution of the atmosphere in the region, a quasi-homogeneity of the obtained AOD fields should be expected; and the appearance of inhomogeneities, most probably, should be caused by the errors of AOD retrieval method, arising due to the neglect of the contribution of the underlying surface.

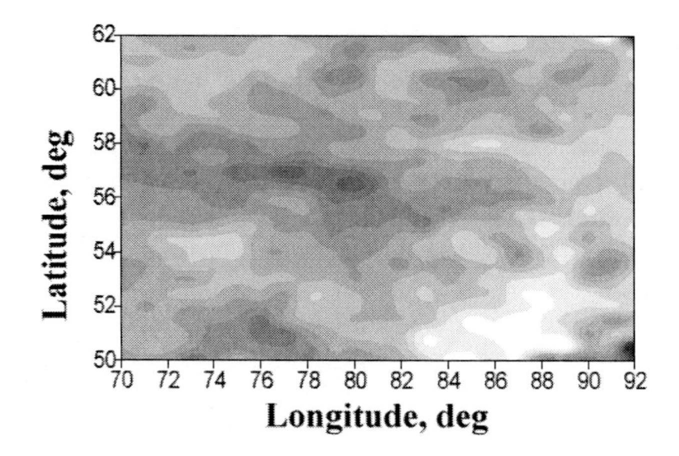

Figure 12. Seasonally average spatial distributions of the aerosol optical depth (λ=0.47 μm); Terra satellite, 2001.

Despite the presence of the spatial AOD inhomogeneities, analysis of these $\tau(x,y)$ for 2001-2004 makes it possible to conclude that the inhomogeneities are within the methodic error of AOD retrieval from satellite data.

Correlation Analysis of Data

At the final stage of the validation of the satellite AOD measurements according to data of work [[49]], we performed an analysis of the coefficients $R_{\mathrm{AOT}}(x,y)$ of the spatial correlation of AOD, which involved studying:

a) the shape of $R_{\mathrm{AOT}}(x,y)$, where x,y are the geographic longitude and latitude;

b) the radial dependence $R_{\mathrm{AOT}}(d)$, where d is the separation of the point of satellite observations from the central point (Tomsk); and

c) the azimuthal dependence $R_{\mathrm{AOT}}(\varphi)$, where φ is the azimuth measured from the north direction.

The studies were performed using the satellite information database MODIS Aerosol Products (Level 2, Collection 005) for the period 2001-2006 and for the region with the coordinates 50-64°N and 65-105°E. During solution of the problems, posed in the work, we had processed about 10000 files MOD04_L2 (Terra satellite, data for 2001-2006) and MYD04_L2 (Aqua satellite, data for 2003-2006), obtained via Internet from *Goddard Distributed Active Archive Center* (DAAC), NASA. For each of the considered year, we selected AOD data pertaining to the seasonal period May-September, when the state of the surface permits satellite-based monitoring of the atmospheric aerosol. The seasonally average characteristics were spatially averaged for the region 50-64° N, 65-105° E, which was divided into surface areas with size of 0.5° latitude by 1° longitude. Thus, the nominal spatial resolution (at nadir) of the averaged data is 56×62 km (for the geographic latitude of Tomsk).

The coefficients of the spatial correlation of the aerosol optical depth $R_{AOT}(x,y)$ were obtained within radius 1100 km of the central point with the geographic coordinates 56.5°N and 85.0°E (Tomsk). To study the azimuthal dependence of the coefficients $R_{AOT}(\varphi)$, data were selected in four angular sectors (Tomsk being the central point) within ±45° of the north ($\varphi=0°$), east ($\varphi=90°$), south ($\varphi=180°$), and west ($\varphi=270°$) directions.

For the spectral channel $\lambda=0.47$ μm, Figure 13 presents the $R_{AOT}(x,y)$ maps, obtained according to MODIS data onboard Terra satellite for period 2001-2006. The $R_{AOT}(x,y)$ maps are complemented with data on the radial dependence of the function $R_{AOT}(d)$. From analysis of the obtained data we can conclude that the configurations of the $R_{AOT}(x,y)$ fields show interannual diversity in the regions of the positive AOD correlation; therefore, two types of the $R_{AOT}(x,y)$ fields can be introduced.

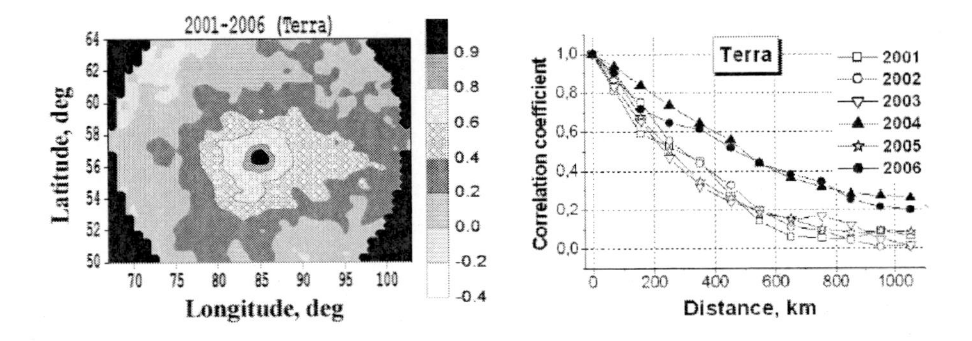

Figure 13. Coefficients of the spatial correlation of AOD according to MODIS data: Terra satellite, $\lambda=0.47$ μm.

First type of R_{AOT} (data for 2001, 2003, and 2005) includes the "spatially localized" fields. They are characterized by relatively rapid decrease of the $R_{AOT}(d)$ values with a growth of the parameter d, when at the distance $d=400$ km the azimuthally average R_{AOT} value is no more than 0.4, while for $d=600$

km the R_{AOT} value is less than 0.2. The second type of R_{AOT} (data for 2002, 2004, and 2006) includes the "spatially blurred" fields. Compared to the first type, these latter are characterized by a slower change of the $R_{AOT}(d)$ values. In this case, the azimuthally average correlation coefficient exceeds R_{AOT} for the first type by as much as 0.2

As regards the azimuthal orientation of R_{AOT}, data of Fig. 13 suggest that the "east-west" axis, on the average, is dominated by the east direction, when the spatial R_{AOT} field is elongated in the east direction, and $R_{AOT}(\varphi=90°)$ values are larger than $R_{AOT}(\varphi=270°)$.

To estimate the adequacy of the obtained correlation fields, we preformed the calculations of the coefficients R_{AOT} according to AERONET data, and compared these results in Table 11 with the data, obtained according to the satellite measurements.

Table 11. Coefficients of the spatial correlation of AOD (λ=0.47 μm)

AERONET stations	Year	AERONET	MODIS
Kireevsk	2001-2006	+0.87	+0.89
Krasnoyarsk	2004	+0.87	+0.83
Irkutsk	2005	-0.24	-0.15

Summarizing the complex validation of the satellite AOD measurements, we should conclude that the satellite AOD measurements have an error of the order of 15-20%, meeting the accuracy requirements of the RTM method, and, moreover, they have adequate spatial and correlation characteristics (Table 10).

3.4. Testing of RTM Method for Meteorological Conditions of Tomsk

In the concluding part of section 3, we performed a joint validation of the RTM method [[30]] and the regression SW algorithm [[21],[22]] for the meteorological conditions of Tomsk. It should be noted that, among the publications concerning the validation of the SW algorithm, some works [[50]-[52]] obtained an accuracy better than 1 K, while others [[53],[54]] reported that the error of the algorithm reached 1.5-3 K.

- A general approach to solving the posed problem was to compare the results of LST retrieval from data of the EOS/MODIS satellite system against the results of the soil temperature measurements, which are regularly conducted at the meteorological site in Tomsk (coordinates are 56°26.5' N, 84°58.7' E). The site location is symbolized by **A** on the map of this region of Tomsk (Fig. 14), obtained with the help of Google Earth.

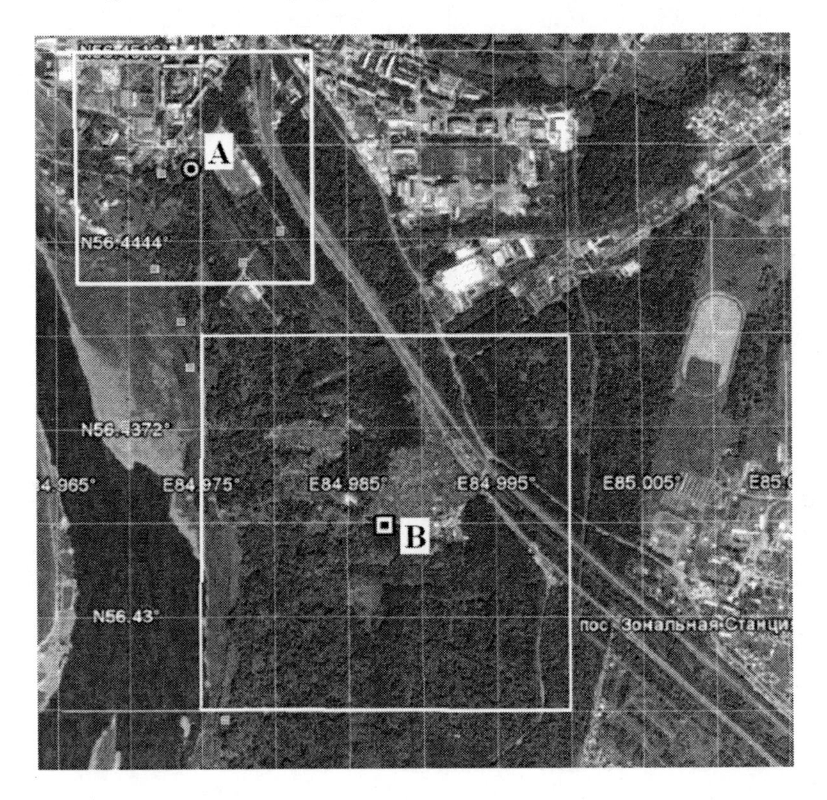

Figure 14. Image of the region of the test works (according to data of Google Earth), symbol A denotes the center of the meteorological site (1×1 km), and symbol B denotes the center of the test area (1.5×1.5 km).

To solve the posed problem, we used the following initial data:

- results of the ground-based measurements of soil temperature for the period from May to September 2001-2004;

- files with EOS/MODIS telemetric information for the Terra satellite, obtained from LAADS Web website at http://ladsweb.nascom.nasa.gov/data/search.html;
- results of LST retrieval in the IR channels 22, 29, 31, and 32 of EOS/MODIS on the basis of thematic processing of the MODIS telemetry information with the use of the RTM method and RTM-based computer program (see section 3.1);
- files with results of LST retrieval from MODIS satellite measurements on the basis of the regression SW algorithm (MOD11_L2 v.5), obtained from LP DAAC Web website at https://lpdaac.usgs.gov/lpdaac/get_data/data_pool.

The satellite LST measurements were selected, as in work [[53]], only for nighttime images. In this case, we can expect much weaker spatiotemporal variations in the temperature of the test surface area than from daytime images. Here, there may not be situations when the temperatures of the shadowed and sunlit portions of the surface differ by 20 K.

The maximum time non-coincidence between the satellite and ground-based measurements did not exceed 30 min, and the selected MODIS images fell within the time interval 20:30–21:30 LT. In addition, using Cloud Mask (MOD35), the satellite images were divided into two categories: cloudless (CM=3) and cloudy (CM=2,1) images. The statistical sample of satellite and ground-based soil temperature measurements, obtained after this selection, consisted of 95 pairs for the cloud-free observation conditions and 27 pairs for the cloudy situations.

For the comparative analysis of the MODIS data (nominal spatial resolution of 1000 m) and local ground-based soil temperature measurements, it is necessary that the test area, with an account for the error in the geographic referencing of the satellite data, possess the relative spatial homogeneity within 1-2 satellite image's pixels. It should be noted that the meteorological site is located in an immediate vicinity of the river and urban buildings. This leads to the fact that within one satellite image's pixel there are objects whose temperatures markedly differ from the temperature measured at the meteorological site. In this regard, at the first stage of the works, for the test area **A** we studied the spatial distribution of the surface temperature. For this, we used high spatial resolution satellite images which were obtained for cloud-free conditions from Landsat 7 satellite in the thermal channel of the ETM+ instrument (a linear pixel size is 57 m). According to the ETM+ data, for the test area **M**, the standard deviation of the satellite-measured radiation temperatures reaches 2.5 K. Therefore, based on these same satellite data, we studied the territory adjacent to the test area **A,** and selected a new test area about 2 km away from the meteorological site. The center of this area is denoted by the symbol **B** in Fig. 14**.** The linear size of the new area exceeded 1500 m, the standard deviation of the radiation temperature had now been less than 1 K, and the average temperature of the area **B** was close to the temperature at the meteorological site.

Figure 15 (upper panel) presents the results of comparative analysis of the data of ground-based LST measurements (T_S) and the values of the surface temperature, retrieved according to the MODIS satellite data with the help of the regression SW algorithm (T_{SW}) and the RTM method (T_{RTM}) for the cloud-free observation conditions. Based on analysis of these data, we can note that the ground-based and satellite measurements have high correlation coefficients R=0.94-0.95 for both methods and close standard deviations, less than 1.5 K. Noteworthy, for the spectral algorithm the average value of $\delta T_{SW}=T_S-T_{SW}$ is about μ_{SW}=+1.4 K; while the average value of $\delta T_{RTM}=T_S-T_{RTM}$ does not exceed μ_{RTM} =-0.2 K. Thus, the obtained result agrees with conclusions, made in works [[53],[54]], that the regression algorithm [[21],[22]] may underestimate the retrieved LST values.

For the cloudy observation conditions, the results of the comparative analysis of ground-based and satellite measurements are presented in Fig. 15 (lower panel). According to these results, in the presence of clouds the μ_{SW} value has increased by 0.9 K, and has now been +2.3 K; while the μ_{RTM} value changed by just -0.2 K, up to the level -0.4 K. At the same time, the correlation coefficient has decreased to R=0.90 for the spectral algorithm and remained the same (R=0.95) for the RTM method. That is, under the cloudy observation conditions, the uncertainty of the SW algorithm has markedly increased; while the accuracy of the RTM method has been only slightly worse than for the cloud-free observation conditions.

Summarizing, we can conclude the following. The results, obtained in this work, and dealing with validation of two methods of the atmospheric correction of the satellite LST measurements, demonstrate clear advantages of our physical (RTM) approach, which shows higher accuracy than the regression method, both for conditions of clear-sky measurements, and in the presence of clouds. Taking into account the uncertainty of the test data on the soil temperature (of the order of 0.5 K), the root-mean-square error of the RTM method was about 1 K for both types of the meteorological conditions.

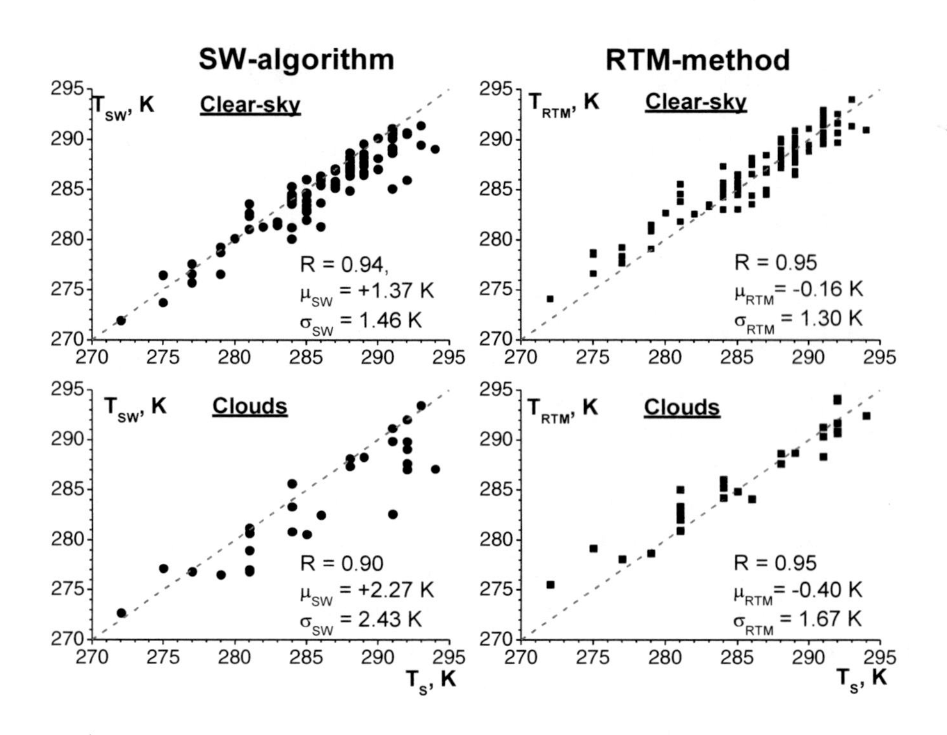

Figure 15. Results of testing two satellite methods for meteorological conditions of Tomsk: R is the correlation coefficient, and μ_{SW} and μ_{RTM} are the average temperature discrepancies.

Chapter 4

APPLICATIONS OF RTM-METHOD

4.1. AN EXAMPLE OF THE RTM-METHOD APPLICATION

To test preliminarily the program complex performance, the temperature of some area of 60×60 km of the Luginetsk oil-gas condensate field (58.15° N, 78.89° E) was sensed under different atmospheric conditions.

A) Cloudless atmosphere (06/02/2004; 12:38 LT)

B) Smoke, cloudiness (06/05/2004; 13:15 LT)

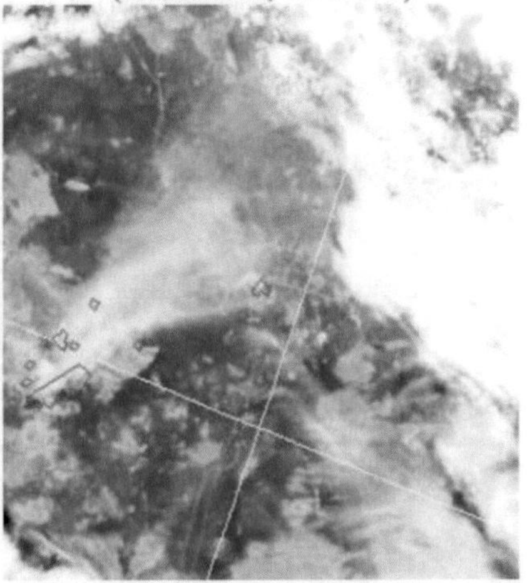

Figure 16. MODIS space images of the Luginetsk oil-gas condensate field; geographical projection – Albers Conical Equal Area.

Figure 16 shows space-born patterns of the area with spatial resolution of 250 m, resulted from the composition of three spectral channels of the visible range of EOS/MODIS space-born system for June 2 and 5, 2004. Torches of the oil-gas condensate field are located at the image center.

In the image A, where the atmosphere is free of clouds and the aerosol has the background content, two basic types of surface are clearly seen: areas covered with vegetation (dark) and open areas (light). Analysis of a series of the cloud-free images allows us to conclude that the spatial distribution of the underlying surface temperature follows the image outlines and is sufficiently stable, as well as that the temperature of light areas is 2–3 K higher than that of dark ones.

A different situation is seen in the image B: the smoke of the forest fires, as well as dense and semi-transparent cloudiness noticeably distort spatial outlines of the underlying surface.

Space-born images in Fig. 16 are supplemented by data in Fig. 17, which are given in Cartesian coordinates.

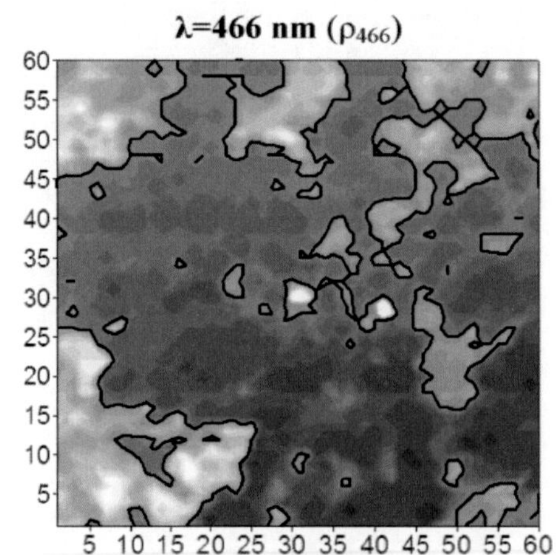

contour: $\rho_{466} = 0{,}1$

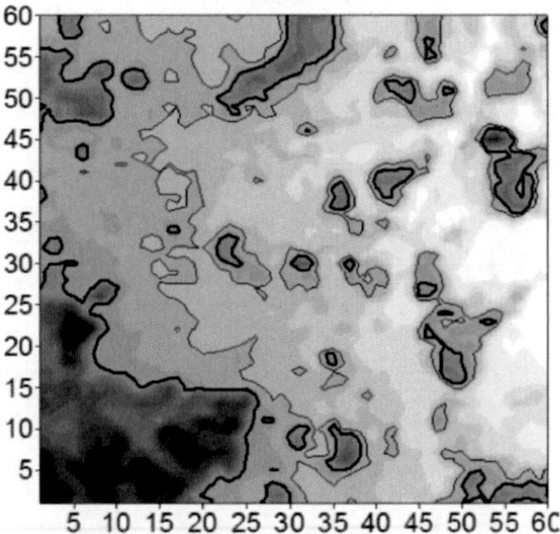

contours: T_{11} = 294.5 and 295.5 K

a

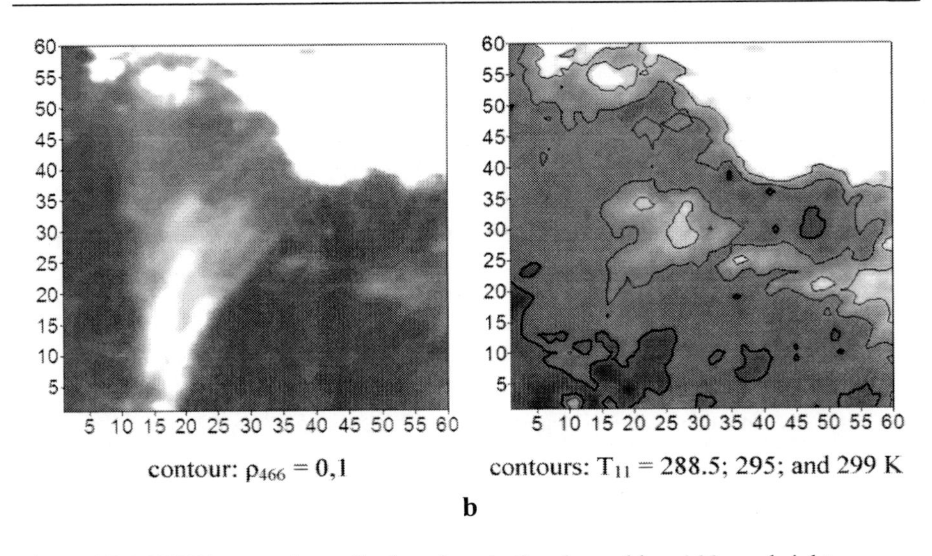

contour: $\rho_{466} = 0{,}1$ contours: T_{11} = 288.5; 295; and 299 K

b

Figure 17. MODIS space data: albedo values in the channel λ = 466 nm, brightness temperatures in the channel λ = 11 μm.

For the case A, open areas are outlined; the characteristic values of albedo measurements at λ=466 nm (ρ_{466}) and the brightness temperature at λ=11 μm (T_{11}) are marked. Bright torches (λ=466 nm) and the sand quarry (to the right) are well seen at the image center. The differences of brightness temperatures in the channels λ=11 and 12 μm characterize the scale of spatial dissimilarity of the distorting atmospheric properties in the IR spectral range.

Analysis of the values $T_{11} - T_{12}$ for the case A leads to a conclusion that distorting properties of the atmosphere in this case can be considered quasihomogeneous (one range of values 0.5–1.3 K, r.m.s. = 0.11 K). In contrast, in the case B, the values of $T_{11} - T_{12}$ fall in the range 0.4–7.1 K and r.m.s. = 0.71 K due to the smoke and cloudiness.

Naturally, the results of operation of the basic SW algorithm [[21],[22]] are of interest (a standard product is called MOD11_L2) in the both cases under consideration. These data, obtained from the site Land Processes Distributed Active Archive Center (LP DAAC, http://edcdaac.usgs.gov/datapool/datapool.asp), are presented in Fig. 18.

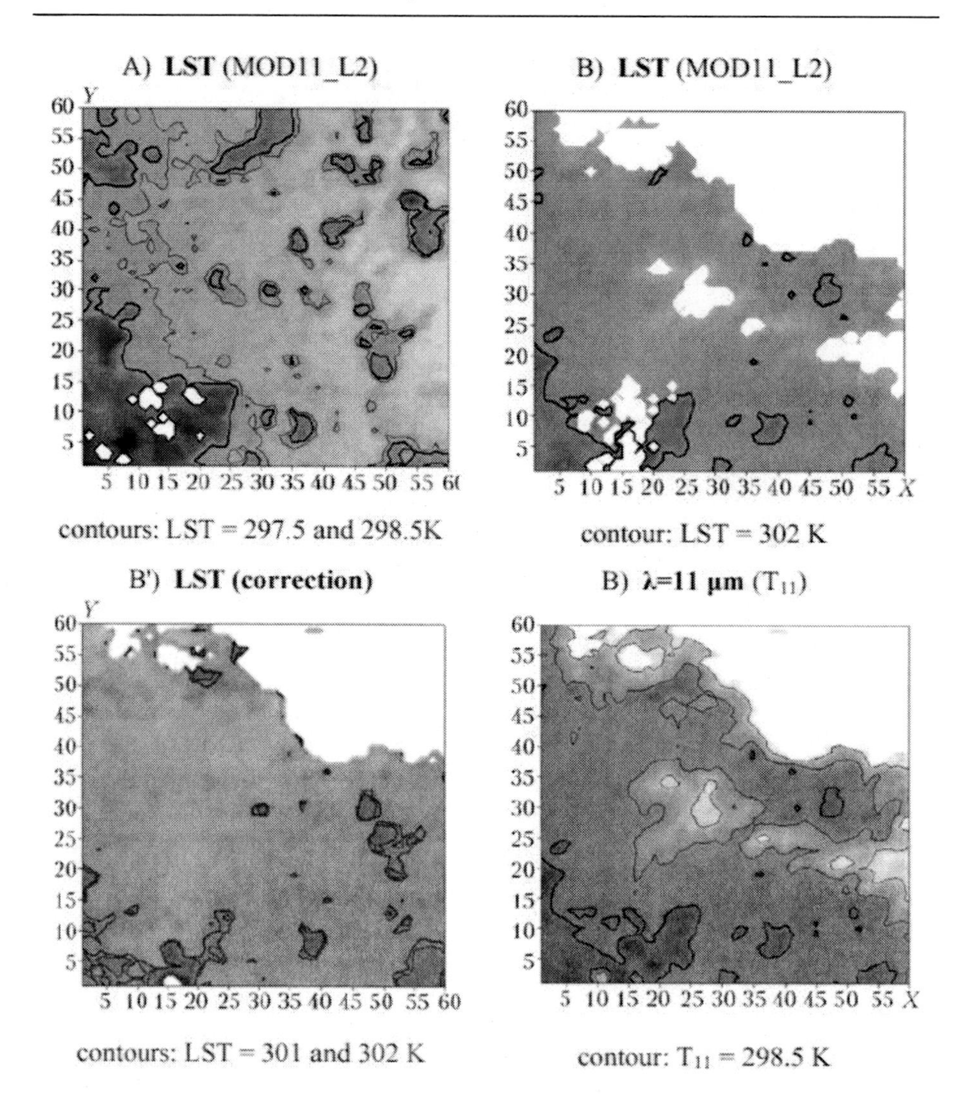

Figure 18. Results of the LST retrieval by the SW algorithm (two top images) and based on the RTM-method.

In case of the cloudless atmosphere, the standard algorithm MOD11 retrieves the LST everywhere, excluding only a few pixels, because the cloud mask, due to bright pixels, fixes (erroneously, to our opinion) the presence of partial cloudiness. In this case, LST spatial structure is almost similar to the spatial structure of brightness temperatures.

In the case B, spatial structures of ρ_{466}, T_{11}, and LST are noticeably distorted by the smoke and cloudiness. The number of white-colored pixels, where MOD11 data are absent, significantly increased. The retained T_{11} structures are outlined in the image. Note that the results of LST retrieval in this case, aside from misses, contain explicitly underestimated LST values, falling within the limits of cloud outlines.

Now we have to fill white-colored misses in LST in order to retrieve the temperature structure similarly to the case of the cloudless atmosphere through the atmospheric correction of data at the areas, where the thermal radiation passes mostly through the smoke and aerosol.

The designed program complex was used for this purpose. As the *a priori* information, we used the IMAPP-retrieved data from MODIS measurements. Using the MODTRAN program in [[30]], the distorting characteristics of the atmosphere were calculated for the channel λ=11 μm (the bottom panels in Fig. 18), measurements of the brightness temperatures T_{11} were corrected, and new spatial distribution B′ of LST was found.

After the correction, a part of the temperature structure of the area, centered at the point (Y = 25, X = 52), was retrieved. Note an important fact that the difference between the retrieved LST values for B and B′ in the vicinity of this point does not exceed 1 K. Besides, the surface temperature near torches (Y = X = 30) was also retrieved.

Consider one more result of the program complex application, i.e., the retrieval of the thermal brightness temperature of torches T_F from measurements of brightness temperatures T_4 in the channel λ = 3.96 μm, accounting for (based on IMAPP data) the optical-meteorological state of the atmosphere [[40]]:

$$B(T_\lambda) = I_F + I_{BG},$$
$$B_F = p(\theta)\varepsilon_\lambda{}^F B(T_F)P_\lambda;\ I_{BG} = I_{SRF} + I_{ATM} + I_{RFL} + I_{SCT},$$
$$B(T_F) = [B(T_\lambda) - I_{BG}] \,/\, [p(\theta)\varepsilon_\lambda{}^F P_\lambda],$$

where $B(T_\lambda)$ is the Planck function; T_λ is the brightness temperature of the thermal radiation; I_F is the intensity of torch emission attenuated by the atmosphere; I_{BG} is the background radiation intensity; I_{SRF} is the contribution of the surface thermal radiation attenuated by the atmosphere; I_{ATM} is the contribution of thermal radiation of the atmosphere; I_{RFL} is the contribution of the incident fluxes of thermal and solar radiation, reflected from the surface; I_{SCT} is along-path scattered (thermal and solar) radiance; $P_\lambda = \exp\{-\tau_\lambda\}$ is the atmospheric transmittance; τ_λ is the optical thickness of the atmosphere; $S(\theta)$ is the ratio of the torch area to pixel size.

Table 12 illustrates the results of the problem solution provided the flame diameter is 19 m.

Table 12.

Case	LST, K	P_λ, τ_λ	T_{11}, K	T_4, K	T_F, K
A	300.0	0.822, 0.196	297.4	334.0	1252
B	302.0*	0.344, 1.067	293.8	321.0	1261

* LST = $T_{11,cor}$ is the temperature retrieved after atmospheric correction of measurements in the channel $\lambda = 11$ μm with the help of the developed program complex (see Fig. 18, case B′).

Thus, retrieving results for torch flame temperature are very close despite different atmospheric conditions during space-born observations. It should be noted that excluding this operation (atmospheric correction) results in significantly different values for A and B cases:. 1192 K и 956 K, i.e., the overestimation for the case B exceeds 300 K.

4.2. Application of the RTM Method to Detection of High-Temperature Objects

The RTM method was tested using data of 97 files (granules) of the telemetric information from EOS/MODIS (Terra satellite, daytime images) for June 2006, pertaining to the West Siberian territory. As test objects for observations, we have chosen 13 torches from combustion of accompanying gas in oil-gas fields of Tomsk and southern Tyumen Regions.

The choice of torches was determined by their stability and availability of their geographic coordinates, necessary for the torch identification. Thus, the torches were a set of varying-intensity thermal objects, allowing the effective elaboration of the methods of satellite-based HTO detection under different conditions of satellite observations.

For elaboration of satellite methods, we used two variants of the standard algorithm MOD14_v5.0.1 [[12]], as well as the RTM methods with the use of our methodical innovations and software [[30]].

To increase the sensitivity of this algorithm in detection of HTOs with a relatively low intensity of the thermal emission, we have modified the MOD14 algorithm by:

a) considerably lowering the thresholds (5): $T_4 > 302$ K (versus former 310 K) and $\Delta T > 3.5$ K (versus 10 K); and

b) changing the coefficients $C_1 \ldots C_4$ (6): $C_1 = 2.5$, $C_2 = 5.0$, $C_3 = 2.0$.

4.2.1. Description of the Algorithm Based on the RTM Method

Stage 1. Based on the EOS/MODIS satellite telemetry, the IMAPP program is used to determine the *a priori* optical-meteorological information on the atmosphere state for regions of detecting the high-temperature sources. The *a priori* information includes the following data:

- a spatial resolution of 1 km: the cloud mask (MOD35), the integrated atmospheric moisture content (MOD05);
- a spatial resolution of 5 km: vertical profiles of the geopotential, the air temperature, humidity, ozone content (MOD07), and cloud characteristics (MOD06);
- a spatial resolution of 10 km: aerosol optical characteristics (MOD04).

emissivities of the pixels ε_λ are determined by the standard method based on maps of surface types and tables of the correspondence of ε_λ to these Earth's surface types.

Stage 2. The cases of water pixels, as well as pixels, covered with thick clouds, are rejected with the use of MOD35, MOD06, and MOD05 data.

Stage 3. For channels 21/22 (henceforth, channel 21), 31, and 32, the *a priori* information, obtained earlier, is used to calculate the characteristics of the thermal radiation distortion by means of the modified version of MODTRAN_v3.5 program. Then, based on the solution of thermal radiative transfer equation, $T_{S,21}$, $T_{S,31}$, and $T_{S,32}$ are calculated, i.e., LST values, retrieved in channels 21, 31, and 32. For correct temperature and humidity profiles in the absence of LST values, the condition of approximate equality $T_{S,21} \approx T_{S,31} \approx T_{S,32}$ is to be satisfied.

Stage 4. If $T_{S,31} \neq T_{S,32}$, then one of the reasons for this mismatch could be errors in profiles of the meteorological parameters. In this case, the simplest compensation for these errors is performed by calculating corrections of the form $\Delta T_S = C_{ERR}(T_{S,32} - T_{S,31})$ and by calculating a new value of $T_{S,31} = T_{S,31} - \Delta T_S$.

Stage 5. In the case of influence of cirrus and semitransparent clouds, the retrieved LST values are corrected: $T_{S,21} = T_{S,21} + \Delta T_{21,CLD}$, $T_{S,31} = T_{S,31} + \Delta T_{31,CLD}$, where the "cloud" corrections are determined via *Look-Up-Table* of the influence of cloud characteristics on LST retrieval results and the mutual analysis of MOD35, MOD06, and MOD05 data.

Stage 6. The HTO detection is performed with the use of two conditions:

$$T_{S,21} > 302 \text{ K and } \Delta T = T_{S,21} - T_{S,31} > 3.5 \text{ K}.$$

4.2.2. Detection Results

Table 12 presents the results of detection of test objects (torches) with the use of two (original and author-modified) MOD14 algorithms, as well as the RTM method for the temperature monitoring of the Earth surface, proposed by us.

Table 13 gives results of torch detection, summed over all torches (N_Σ), the number of detections of each torch, and average temperature for each torch ($T_{21,av}$).

For the algorithm testing, a total of 38 128 pixels in the torch neighborhood were processed. Note that the condition $T_{S,21} \approx T_{S,31} \approx T_{S,32}$ in the absence of clouds and HTOs does hold, signifying a good quality of the atmospheric correction of satellite LST measurements. For instance, for the sample, consisting of 30985 pixels, corresponding to conditions of the clear-sky atmosphere, average retrieved LSTs were: $T_{S,21} = 298.4$ K, $T_{S,31} = 298.4$ K, $T_{S,32} = 298.7$ K. That is, the uncertainty in accounting for the molecular absorption in the EOS/MODIS channels 21/22, 31, and 32 was, on the average, less than 0.5 K.

The number of torch detections N_Σ with the use of the MOD14 v5.0.1 algorithm was 60, with identification of 6 test objects out of 13. For the MOD14_v5.0.1 algorithm, modified by us, (MOD14*), $N_\Sigma = 83$ with 10 test objects identified. With the use of the RTM method, N_Σ reached 122, and all 13 test objects were observed at a varying frequency.

Table 13. Results of detection of 13 test objects (torches) from space with the help of three satellite methods

Method	N_Σ	Torches												
		F1	F2	F3	F4	F5	F6	F7	F8	F9	F10	X1	X2	X3
MOD14	60	4	–	–	–	–	–	–	–	1	1	14	14	26
MOD14*	83	6	2	–	1	–	1	1	–	6	4	18	18	26
RTM	122	13	4	3	4	2	8	1	1	8	9	21	21	27
T_{21}		309	304	306	306	305	305	308	303	307	306	314	320	329

Thus, the RTM method is, on the average, a factor of two more efficient than the standard MOD14_v5.0.1 algorithm. In the modified algorithm version MOD14*, the detection thresholds of potential fires coincide with thresholds in the RTM method. However, in this case again the RTM method is much (almost a factor of 1.5) more efficient than MOD14*.

Speaking about comparative estimates of the efficiency of these three algorithms, it is very important to note the following. Among test objects we can distinguish three bright torches (X1...X3, see Table 13), located in the south of the Tyumen Region, for which the detection frequency is markedly higher than for other torches at a less dependence on the choice of the method. Considering that the RTM method shows its main advantages in detection of relatively low-intensity thermal sources, it is advisable to obtain comparative estimates of application of the methods to such sources, namely, ten torches (F1...F10, see Table 13), located in Tomsk Region. In this case, N_Σ ratio for the three considered algorithms is already 6:21:53 for MOD14, MOD14*, and RTM, respectively, therefore, advantages of the RTM method markedly increase.

Let us compare the efficiency of application of the RTM method and the algorithm used at IAO SB RAS [[6],[7]], namely, the algorithm of the forest fire detection from data of the NOAA POES satellite system, when detecting low-intensity torches. In this case, the N_Σ ratio will be 36:53 for the IAO algorithm and the RTM method, respectively. In detection of high-temperature objects, the RTM method has considerable advantages over standard approaches, especially, for the problem of detection of weak-intensity sources under complex optical-meteorological observation conditions.

Thus, among the considered satellite methods, the RTM method, proposed by us, is the most efficient. Then the IAO algorithm follows next, and two variants of the MOD14 algorithm conclude the list.

Chapter 5

CONCLUSION

The main results and data of satellite monitoring of boreal forest fires on the territory of the Tomsk Region as follows.

1. The effectiveness of application of the AVHRR/NOAA satellite system for satellite monitoring of forest fires on this territory is in the range 19–48% (about 38% on average) and depends on the seasonal characteristics of fires themselves (size and time of burning) and state of cloudiness.

2. The probability of early fire detection from satellites is in the range 13–27% (about 18% on average).

3. The minimal sizes of the forest fires fixed in the regime of automatic AVHRR data interpretation are about 0.1–0.2 ha, and they are detected with probability of about 10%. For the limiting area of effective fire extinguishing (about 5 ha), the detection probability increases to 35–45%, which suggests the possibility of efficient application of satellite systems for early detection of fires in their early stages.

4. The efficiency of the regional algorithm developed at the IAO SB RAS is much greater than that of the MODIS Fire Product algorithm.

5. The maximum efficiency of satellite monitoring of forest fires is offered only by the complete SMFF scheme, including satellite image recording irrespective of the time of day.

6. The RTM method based on real-time satellite meteorological data on the atmospheric state at the moment of satellite observation allows the distorting effect of the molecular atmosphere to be considered with errors less than 0.5 K. Application of the RTM method based on the *split-window* principle makes

this solution stable in the terms of errors of assigning *a priori* meteorological information.

7. The RTM method has significant advantages over the standard approaches in detecting high-temperature sources, especially low-intensive fire sites under unfavorable meteorological conditions of observations.

8. The software prototype for real-time correction of satellite IR MODIS measurements for the distorting effect of the atmosphere presented in this work allows the capabilities of the existing methods of temperature sensing of the underlying surface to be extended through accounting for the distorting effect of the aerosol and semitransparent cloudiness.

ACKNOWLEDGMENTS

In conclusion we would like to express our gratitude to *managers and staff management in* Forest Protection Services *for a fruitful and long-term cooperation. We are grateful to former* deputy director, Prof., V.V. Koshelev (Institute of Solar-Terrestrial Physics SB RAS), deputy director, Prof., E.A. Loupian (Space Research Institute RAS) and Dr. A.I. Sukhinin (Sukachev Institute of Forest SB RAS) for useful advice and discussions. We are thankful to our colleagues Yu.V. Gridnev, Dr. D.V. Solomatov, M.V. Engel′, N.V. Kabanova (Laboratory of Optical Signal Propagation, IAO SB RAS) for a *great contribution to the detection of fires.* We are also thankful to V.P. Protasova for help in preparation of this book.

REFERENCES

[1] Zuev, V.E.; Selivanov, A.S.; Fomin, V.V.; Panfilov, A.S.; Romanov, A.V.; Afonin, S.V.; Khamarin, V.I. Measuring the ocean surface temperature with the MSU-SK sensor from the Kosmos-1689 satellite. *Atmos. Oceanic Opt.* 1988, *1*, 76-80.

[2] Zuev, V.E.; Belov, V.V.; Veretennikov, V.V. *Systems with Applications in Scattering Media*. Publishing House of the SB RAS: Tomsk, RU, 1997; 402 pp. (in Russian).

[3] Afonin, S.V.; Panfilov, A.S.; Romanov, A.V.; Selivanov, A.S.; Fomin, V.V.; Khamarin, V.I. Satellite experiments on testing radiometric accuracy of IR-channels of the MSU-SK sensor placed onboard Resurs Satellite No.2 during LKI. *Proc. SPU «Planeta»*. 1993, *42*, 33-37 (in Russian).

[4] Belov, V.V., Afonin, S.V., and Makushkina, I.Yu. Image transfer through the atmosphere. *Atmos. Oceanic Optics*. 1997, *10*, 449-462.

[5] Belov, V.V., Afonin, S.V., Gridnev, Yu.V., and Protasov, K.T. Thematic processing and atmospheric correction of satellite images. *Atmos. Oceanic Opt*. 1999, *12*, 991-997.

[6] Afonin, S.V.; Belov, V.V.; Gridnev, Yu.V. System of the space-based monitoring of forest fires on the territory of Tomsk Region. Part 1. Organization of the space-based monitoring system. *Atmos. Oceanic Opt*. 2000, *13*, 921-929.

[7] Afonin, S.V.; Belov, V.V. System of the space-based monitoring of forest fires on the territory of Tomsk Region. Part 2. Estimation of efficiency of space monitoring. *Atmos. Oceanic Opt*. 2001, *14*, 634-638.

[8] Kaufman, Y.J.; Justice, C. *MODIS ATBD: Fire Products (Version 2.2 Nov.10 1998)*. EOS ID#2741. 1998, 77 pp.

[9] Boles, S.H. and Verbyla, D.L. Comparison of Three AVHRR-based Fire Detection Algorithms for Interior Alaska. *Remote Sens. Environ.* 2000, *72*, 1–16.

[10] Li, Z.Q.; Nadon, S.; Cihlar, J. Satellite-based detection of Canadian boreal forest fires: development and application of the algorithm. *Int. J. Remote Sens.* 2000, *21*, 3057-3069.

[11] Li, Z.; Kaufman, Y.J.; Ichoku, C.; Fraser, R.; Trischenko, A.; Giglio, L.; Jin, J.; Yu, X. A review of AVHRR-based active fire detection algorithms: principles, limitations, and recommendations. In F. Ahern, J. Goldammer, C. O. Justice (Eds.), *Global and regional wildfire monitoring from space: Planning a coordinated international effort.* SPB Academic Publishing: Hague, NE, 2001; 199-225.

[12] Giglio, L.; Descloitres, J.; Justice, C.; Kaufmann, Y. An Enhanced Contextual Fire Detection Algorithm for MODIS. *Remote Sens. Environ.* 2003, *87*, 273-282.

[13] Loupian, E.A.; Mazurov, A.A.; Flitman, E.V.; Ershov, D.V.; Korovin, G.N.; Novik, V.P.; Abushenko, N.A.; Altyntsev, D.A.; Koshelev, V.V.; Tashchilin, S.A.; Tatarnikov, A.V.; Csiszar, I.; Sukhinin, A.I.; Ponomarev, E.I.; Afonin, S.V.; Belov, V.V.; Matvienko, G.G.; Loboda T. Satellite monitoring of forest fires in Russia at federal and regional levels. *Mitigation and Adaptation Strategies for Global Change*. 2006, *11*, 113-145.

[14] Franca, J.R.de A.; Brustet, J.-M.; Fontan, J. Multispectral remote sensing of biomass burning in West Africa. *J. Atmos. Chem.* 1995, *22*, 81–110.

[15] Minko, N.P.; Abushenko, N.A.; Koshelev, V.V. Forest fire detection in East Siberia forests using AVHRR/NOAA data. *Proceedings of SPIE.* 1998, *3502*, 192-200.

[16] Gridnev, Yu.V. Detection of small fires from NOAA/AVHRR data. *Atmos. Oceanic Optics*. 2002, *15*, 659-662.

[17] Becker, F.; Li, Z.L. Towards a local split window method over land surface. *Int. J. Remote Sens.* 1990, *11*, 369-393.

[18] Ottle', C.; Vidal-Madjar, D. Estimation of land surface temperature with NOAA 9 data. *Remote Sens. Environ.* 1992, *40*, 27-41.

[19] Li, Z.L.; Becker, F. Feasibility of land surface temperature and emissivity determination from AVHRR data. *Remote Sens. Environ.* 1993, *43*, 67-85.

[20] Sobrino, J.A.; Li, Z.L.; Stoll, M.P.; Becker, F. Improvements in the split window technique for land surface temperature determination. *IEEE Trans. Geosci. Remote Sens.* 1994, *32*, 243-253.

[21] Wan, Z.; Dozier, J. A generalized split-window algorithm for retrieving land surface temperature measurement from space. *IEEE Trans. Geosci. Remote Sens*. 1996, *34*, 892-905.

[22] Wan, Z. (1999). MODIS Land-Surface Temperature Algorithm Theoretical Background Document (LST ATBD), version 3.3. http://modis.gsfc.nasa.gov/data/atbd/atbd_mod11.pdf

[23] Coll, C.; Caselles, V. A split-window algorithm for land surface temperature from advanced very high resolution radiometer data: Validation and algorithm comparison. *J. Geophys. Res*. 1997, *102 (D14)*, 16,697-16,713.

[24] Mao, K.; Qin, Z.; Shi, J.; Gong, P. A practical split-window algorithm for retrieving land surface temperature from MODIS data. *Int. J. Remote Sens*. 2005, *26*, 3181-3204.

[25] Afonin, S.V. *Development and application of the atmospheric radiation model for detection of the ocean temperature from space sensing data*. Cand. Phys.-Math. Sci. Dissert. Tomsk, RU, 1987; 192 pp. (in Russian).

[26] Belov, V.V., Afonin, S.V. *From Physical Foundations, Theory, and Simulation to Thematic Processing of Satellite Images*. Publishing House of IAO SB RAS: Tomsk, RU. 2005; 266 pp. (in Russian).

[27] Thome, K.; Palluconi, F.; Takashima, T.; Masuda, K. Atmospheric correction of ASTER. *IEEE Trans. Geosci. Remote Sens*. 1998, *36*, 1199-1211.

[28] Sobrino, J.A.; Jiménez-Muñoz, J.C.; Paolini, L. Land surface temperature retrieval from LANDSAT TM 5. *Remote Sens. Environ.* 2004, 90, 434-440.

[29] Wang, P.; Karen, Y.L.; Cwik, T.; Green, R. MODTRAN on supercomputers and parallel computers. *Parallel Computing*. 2002, *28*, 53-64.

[30] Afonin, S.V.; Solomatov, D.V. Solution of problems of atmospheric correction of satellite IR measurements accounting for optical-meteorological state of the atmosphere. *Atmos. Oceanic Opt.* 2008, *21*, 125-131.

[31] Clough, S.A.; Shephard, M.W.; Mlawer, E.J.; Delamere, J.S.; Iacono, M.J.; Cady-Pereira, K.; Boukabara, S.; Brown, P.D. Atmospheric radiative transfer modeling: a summary of the AER codes, Short Communication. *J. Quant, Spectrosc. Radiat. Transfer*. 2005, *91*, 233-244.

[32] Rothman, L.S.; Jacquemart, D.; Barbe, A.; Benner, D.C.; Birk, M.; Brown, L.R.; Carleer, M.R.; Chackerian, C., Jr.; Chance, K.; Dana, V.;

Devi, V.M.; Flaud, J.-M.; Gamache, R.R.; Goldman, A.; Hartmann, J.-M.; Jucks, K.W.; Maki, A.G.; Mandin, J.-Y.; Massie, S.T.; Orphal, J.; Perrin, A.; Rinsland, C.P.; Smith, M.A.H.; Tennyson, J.; Tolchenov, R.N.; Toth, R.A.; Auwera Vander, J.; Varanasi, P.; Wagner, G. The HITRAN 2004 Molecular Spectroscopic Database. *J. Quant. Spectrosc. Radiat. Transfer*. 2005, *96*, 139-204.

[33] Mlawer, M.J.; Tobin, D.C.; Clough, S.A. A Revised Perspective on the Water Vapor Continuum: The MT_CKD Model. *J. Quant, Spectrosc. Radiat. Transfer* (in preparation).

[34] Kneizys, F.X.; Abreu, L.W.; Anderson, G.P.; Chetwynd, J.H.; Shettle, E.P.; Berk, A.; Bernstein, L.S.; Robertson, D.C.; Acharya, P.; Rothman, L.S.; Selby, J.E.A.; Gallery, W.O; Clough, S.A. *The MODTRAN 2/3 Report and LOWTRAN 7 Model*, Phillips Laboratory, Hanscom AFB contract F19628-91-C-0132 with Ontar Corp. 1996.

[35] Berk, A.; Anderson, G.; Acharya, P.; Hoke, M.; Chetwynd, J.; Bernstein, L.; Shettle, E.; Matthew, M.; Adler-Golden, S. *MODTRAN4 Version 3 Revision 1 User's Manual*, Air Force Res. Lab., Hanscom Air Force Base, Mass. 2003.

[36] Rothman, L.S.; Rinsland, C.P.; Goldman, A.; Massie, S.T.; Edwards, D.P.; Flaud, J.-M.; Perrin, A.; Camy-Peyret, C.; Dana, V.; Mandin, J.Y.; Schröder, J.; McCann, A.; Gamache, R.R.; Wattson, R.B.; Yoshino, K.; Chance, K.V.; Jucks, K.W.; Brown, L.R.; Nemtchinov, V.; Varanasi, P., The HITRAN molecular spectroscopic database and HAWKS (HITRAN atmospheric workstation): 1996 edition. *J. Quant Spectrosc. Radiat. Tran*sfer. 1998, *60*, 665–710.

[37] Clough, S.A.; Kneizys, F.X.; Davies, R.W. Line shape and the water vapor continuum. *Atmos. Res*. 1989, *23*, 229–241.

[38] Seemann, S.W.; Li,, J.; Menzel, W.P.; Gumley, L.E. Operational retrieval of atmospheric temperature, moisture, and ozone from MODIS infrared radiances. *J. Appl. Meteorol*. 2003, *42*, 1072–1091.

[39] Li J., Wolf W., Menzel W. P., Zhang W., Huang H.-L., Achtor T. H. Global Soundings of the Atmosphere from ATOVS Measurements: The Algorithm and Validation. *J. Appl. Meteor.* 2000. *39*. P. 1248- 1268.

[40] Afonin, S.V. To the problem of atmospheric correction of satellite data in space monitoring of small-sized forest fire sources. *Atmos. Oceanic Opt.* 2005, *18*, 299–301.

[41] Afonin S.V., Belov V.V., Solomatov D.V. Solution of problems of the temperature monitoring of the Earth's surface from space on the basis of the RTM method. *Atmos. Oceanic Opt.* 2008, *21*, 1056–1063.

[42] Wan Z., MODIS Land-Surface Temperature Algorithm Theoretical Background Document (LST ATBD), version 3.3 Inst. for Comput. Earth Syst. Sci., Univ. of Calif., Santa Barbara 1999. [*Electronic resource*]: http://modis.gsfc.nasa.gov/data/atbd/atbd_mod11.pdf.

[43] Levy R. C., Remer L. A., Mattoo S., Vermote E. F., Kaufman Y. J. (2007), Second-generation operational algorithm: Retrieval of aerosol properties over land from inversion of Moderate Resolution Imaging Spectroradiometer spectral reflectance. *J. Geophys. Res*. 2007. 112, D13211, doi:10.1029/2006JD007811.

[44] Ichoku C., Chu D. A., Mattoo S., Kaufman Y.J., Remer L.A., Tanré D., Slutsker I., Holben B.N. A spatio-temporal approach for global validation and analysis of MODIS aerosol products. *Geophys. Res. Lett*. 2002. V.29, No.12. doi:10.1029/2001GL013206.

[45] Afonin S.V., Engel M.V., Pavlov A.N., Shmirko K.A., Stolyarchuk S.Yu., Bukin O.A. Analysis of spatiotemporal dynamics of aerosol optical depth according to satellite and lidar data in Primorsky Krai during spring 2009 // Proc. of the 25th International Laser Radar Conference (ILRC-25), St.-Petersburg, 5–9 July 2010. P. 500-503.

[46] Afonin S.V., Belov V.V., Engel' M.V. Statistical analysis of the MODIS Atmosphere Products for the Tomsk Region // *Proc. SPIE*. 2005. V. 5979. P. 164-172.

[47] S.V. Afonin, V.V. Belov and M.V. Engel Comparative analysis of space aerosol data of the MODIS Aerosol Products type. *Atmos. Oceanic Opt*. 2008, *21*, 235-239.

[48] Papadimas C.D., Hatzianastassiou N., Mihalopoulos N., Querol X., Vardavas I. Spatial and temporal variability in aerosol properties over the Mediterranean basin based on 6-year (2000-2006) MODIS data. *J. Geophys. Res*. 2008. V. 113. D11205. doi:10.1029/2007JD009189.

[49] Afonin S.V., Belov V.V., Panchenko M.V., Sakerin S.M., Engel' M.V. Correlation analysis of spatial fields of the aerosol optical thickness on the base of MODIS data. *Atmos. Oceanic Opt*. 2008, *21*, 510-515.

[50] Wan Z., Zhang Y., Zhang Q., Li Z.-L. Validation of the land-surface temperature products retrieved from Terra Moderate Resolution Imaging Spectroradiometer data // *Remote Sens. Environ*. 2002. V.83. N 1-2. P. 163-180.

[51] Wan Z., Zhang Y., Zhang Q., Li Z.-L. Quality assessment and validation of the MODIS global land surface temperature // *Int. J. Remote Sens*. 2004. V.25. N 1. P. 261-274.

[52] Wan Z. New refinements and validation of the MODIS land-surface temperature/ emissivity products // *Remote Sens. Environ.* 2008. V.112. N 1. P. 59-74.

[53] Wang W., Shunlin Liang S., Tilden Meyers T. Validating MODIS land surface temperature products using long-term nighttime ground measurements // Remote Sens. Environ. 2008. V.112. N 3. P. 623–635.

[54] Mao K., Shi J., Li Z., Tang H. An RM-NN algorithm for retrieving land surface temperature from EOS/MODIS data. *J. Geophys. Res.* 2007. V. 112. D21102. P. 1-17.

INDEX

A

B

C

D

N

O

P

Q

R

S

T

U

V

W